TECHNICAL REPORT

How Do Earnings Change When Reservists Are Activated?

A Reconciliation of Estimates Derived from Survey and Administrative Data

Francisco Martorell, Jacob Alex Klerman, David S. Loughran

Prepared for the Office of the Secretary of Defense

RAND NATIONAL DEFENSE RESEARCH INSTITUTE

The research described in this report was prepared for the Office of the Secretary of Defense (OSD). The research was conducted in the RAND National Defense Research Institute, a federally funded research and development center sponsored by the OSD, the Joint Staff, the Unified Combatant Commands, the Department of the Navy, the Marine Corps, the defense agencies, and the defense Intelligence Community under Contract W74V8H-06-C-0002.

Library of Congress Cataloging-in-Publication Data is available for this publication.

ISBN 978-0-8330-4474-7

Published 2008 by the RAND Corporation
1776 Main Street, P.O. Box 2138, Santa Monica, CA 90407-2138
1200 South Hayes Street, Arlington, VA 22202-5050
4570 Fifth Avenue, Suite 600, Pittsburgh, PA 15213-2665
RAND URL: http://www.rand.org/
To order RAND documents or to obtain additional information, contact
Distribution Services: Telephone: (310) 451-7002;
Fax: (310) 451-6915; Email: order@rand.org

Preface

This report was produced as part of the RAND project "Activation and the Earnings of Reservists." In related projects, RAND research has shown that, on average, reservists experience large earnings gains while they are activated. These results stand in contrast to estimates derived from the 2004 and 2005 Status of Forces Survey of Reserve Component Members (SOFS-R), which suggest that, on average, reservists suffer large earnings losses while they are activated. This report explores why administrative and SOFS-R data sources produce such divergent estimates of the effect of activation on the earnings of reservists and will be of interest to manpower analysts, survey methodologists, and anyone concerned with the effect of activation on reservists' financial well-being.

The research was sponsored by the Office of the Secretary of Defense (Reserve Affairs) and conducted within the Forces and Resources Policy Center of the RAND National Defense Research Institute (NDRI), a federally funded research and development center sponsored by the Office of the Secretary of Defense, the Joint Staff, the Unified Combatant Commands, the Department of the Navy, the Marine Corps, the defense agencies, and the defense Intelligence Community.

Comments regarding this work are welcome and may be addressed to Paco Martorell at martorell@rand.org. For more information on RAND's Forces and Resources Policy Center, contact the Director, James Hosek. He can be reached by email at James_Hosek@rand.org; by phone at 310-393-0411, extension 7183; or by mail at the RAND Corporation, 1776 Main Street, Santa Monica, California 90407-2138. More information about RAND is available at www.rand.org.

Contents

Preface ... iii
Figures ... vii
Tables ... ix
Summary ... xi
Acknowledgments ... xv
Abbreviations ... xvii

CHAPTER ONE
Introduction ... 1

CHAPTER TWO
Data and Methods ... 3
SOFS-R ... 3
Administrative Data ... 5
Matching the SOFS-R and Administrative Data ... 7
Analysis of Differences in Estimated Earnings Change During Activation ... 8

CHAPTER THREE
Decomposing Differences in Estimated Earnings Changes ... 9
Baseline Difference in Estimates of Earnings Changes ... 9
Difference in Estimates of Earnings Changes Attributable to the Tax Advantage ... 11
Difference in Estimates of Earnings Changes Attributable to Misreported Military Earnings ... 13
Aligning the Military Earnings Concepts ... 13
The Quantitative Importance of Misreported Military Earnings in the SOFS-R ... 13
Explaining the Difference Between the 2004 and 2005 SOFS-R Estimates of Military Earnings ... 16
Difference in Estimates of Earnings Changes Attributable to Civilian Earnings ... 18
Aligning the Civilian Earnings Definitions ... 18
Differences in SOFS-R and Administrative Estimates of Civilian Earnings ... 20
Why Do SOFS-R and Administrative Estimates of Civilian Earnings Differ? ... 21
Explaining the Difference Between the 2004 and 2005 SOFS-R Estimates of Civilian Earnings ... 23

CHAPTER FOUR
Analysis of Nonresponse Bias 27
An Approach to Quantifying Nonresponse Bias in the SOFS-R 27
Estimates of Nonresponse Bias in the SOFS-R 28

CHAPTER FIVE
Conclusion 33

APPENDIX
A. Administrative Data Estimates of Changes in Reserve Earnings Attributable to Activation 35
B. Exact Wording of 2004 and 2005 SOFS-R Earnings Questions 45
C. Detailed Analysis of Differences in Military Earnings 47

Bibliography 53

Figures

3.1. Comparison of 2004 and 2005 SOFS-R Military Earnings Distributions 19
3.2. Comparison of 2004 and 2005 SOFS-R Civilian Earnings Distributions 24

Tables

2.1. SOFS-R and DMDC Administrative Data Match Rates 7
3.1. Estimates of Average Monthly Earnings Change Derived from SOFS-R and Administrative Data 9
3.2. Estimates of Average Monthly Earnings Change Derived from SOFS-R and Administrative Data, Excluding the Tax Advantage 12
3.3. Comparison of Mean Military Earnings in SOFS-R and Administrative Data 14
3.4. Difference in Estimated Earnings Changes Accounted for by Tax Advantage and Misreported Military Earnings 15
3.5. 2004 and 2005 SOFS-R Military Earnings Distributions 20
3.6. Comparison of Mean Civilian Earnings in SOFS-R and Administrative Data 21
3.7. 2004 and 2005 SOFS-R Civilian Earnings Distributions 25
4.1. Pay Grade and Administrative Earnings, by SOFS-R Respondent Status (Unweighted) 29
4.2. Pay Grade and Administrative Earnings, by SOFS-R Respondent Status (Weighted) 30
A.1. Sample Sizes, by Base Year and Out Year and Active-Duty Days Served in the Out Year 37
A.2. Gross and Net Earnings Differences, by Base and Out Year 39
A.3. Gross and Net Earnings Losses, by Base and Out Year 39
A.4. Gross and Net Earnings Differences and Losses, by Number of Active-Duty Days in 2005 39
A.5. Earnings Differences and Losses, by Rank in 2005 40
A.6. Gross and Net Earnings Differences and Losses, by One-Digit Military Occupation 41
A.7. Gross and Net Earnings Differences and Losses, by Three-Digit Military Occupation: Occupations with Earnings Losses Exceeding 20 Percent 42
C.1. Distribution of Difference in Military Earnings, 2004 48
C.2. Distribution of Difference in Military Earnings, 2005 49
C.3. Distribution of Difference in Military Earnings: Basic Pay, 2004 51

Summary

A large fraction of the reserve force has been activated since September 11, 2001, in support of the Global War on Terror and its associated contingencies. Among the hardships of activation is the possibility that the labor market earnings of reservists might fall while they are activated relative to their earnings before being activated. Recent research by Loughran, Klerman, and Martin (2006) (hereafter referred to as LKM) suggests that most reservists in fact earn substantially more while they are activated than they do while not activated. LKM draw this conclusion from administrative earnings records maintained by the Department of Defense (DoD) and the Social Security Administration (SSA). However, self-reported earnings recorded in the 2004 and 2005 Status of Forces Survey of Reserve Component Members (SOFS-R) imply the opposite conclusion: Activated reservists on average experience significant earnings losses.

Estimates of earnings changes derived from SOFS-R and administrative data might differ for a number of reasons. The SOFS-R and administrative data differ in the samples of reservists surveyed, the way earnings are defined, and the time period over which pre- and during-activation earnings comparisons are made. Misreporting and nonresponse bias, problems common to all surveys, might bias estimates derived from the SOFS-R data. On the other hand, civilian earnings may not be recorded perfectly in our administrative data sources, leading to biased estimates derived from those data. In this study, we report on the results of a set of analyses designed to quantify the relative importance of these and other reasons why estimates of earnings changes derived from SOFS-R and administrative data differ.

Matched SOFS-R and Administrative data

Our analyses employ a unique dataset consisting of individual SOFS-R responses matched to administrative data on military and civilian earnings derived from the same sources employed by LKM. When weighted, the 2004 and 2005 SOFS-R were designed to be representative of the Selected Reserves. The surveys record information on a wide range of topics including labor market earnings both before and during activation. The administrative data we used come from a variety of sources. We draw information on military pay from the Active Duty Pay Files and Reserve Pay Files maintained by the Defense Manpower Data Center (DMDC). The pay files contain a detailed breakdown of all compensation that military personnel receive each month and permit the computation of the implicit value of federal income tax exemptions accorded to some military earnings (the federal "tax advantage"). We draw information on civilian earnings from SSA's Master Earnings File (MEF). These SSA earnings records include all earnings subject to Medicare taxes. Although these data cover the vast majority of civilian earnings, they cannot include earnings not reported to SSA, such as any earnings received under the table.

These various datasets were merged with the assistance of DMDC and SSA. RAND supplied DMDC and SSA with programs that analyzed the matched data and generated group-level statistics that could be further processed at RAND without the risk of divulging sensitive survey or SSA earnings data on individuals.

Key Findings

We first established a baseline difference in earnings change estimates. Broadly speaking, the administrative data indicate significant average earning gains whereas the SOFS-R indicates significant average earnings losses. Baseline estimates of monthly earnings changes were $1,665 higher in the administrative data than in the 2004 SOFS-R and $7,247 higher than in the 2005 SOFS-R (the large difference between the 2004 and 2005 SOFS-R results is explained below). We then examined potential explanations for why these sets of estimates differ.

Our analyses depended crucially on our ability to align the definition of earnings in the SOFS-R with the definition of earnings in the administrative data. This alignment was less than perfect for a number of reasons. First, the SSA earnings data are reported on a calendar year basis whereas activation periods frequently span calendar years. Second, the survey does not clearly define the pre-activation period for which respondents are supposed to report earnings. Finally, SSA earnings data do not necessarily record all sources of labor market income, namely, income received "under the table." Because we know that the administrative data record military earnings comprehensively, and because those data are available on a monthly basis, we are more confident in our interpretation of differences in estimates of military earnings across the SOFS-R and administrative data than we are in our interpretation of differences in estimates of civilian earnings across these data sources.

Tax Advantage

The SOFS-R instructs respondents to report pre-tax earnings, but the earnings received by reservists while serving in a combat zone are not subject to federal taxes (or state taxes in some cases). When the implicit value of the federal tax advantage is omitted from the administrative estimates of total earnings, the baseline difference in estimates of earnings changes declines by 28 percent in the case of the 2004 SOFS-R and by 8 percent in the case of the 2005 SOFS-R.

Misreporting of Military Earnings

Military earnings before and during activation are consistently higher in the administrative data than in the 2004 SOFS-R. Because we believe that we can align the military earnings definitions quite closely in the SOFS-R and administrative data, we conclude that respondents in the 2004 SOFS-R, on average, underreport military earnings. Respondents in the 2005 SOFS-R, on average, overreport military earnings. On closer examination, however, the 2005 result is driven by a small number of outliers in the SOFS-R. These comparisons suggest that respondents to the SOFS-R significantly underreport military earnings, especially while activated. This could be because reservists fail to account for the many different types of pays and allowances they receive while serving on active duty.

In the case of the 2004 SOFS-R, we conclude that underreporting military earnings by SOFS-R respondents accounts for up to 42 percent of the baseline difference in estimates of earnings changes. A smaller share of the difference between the 2005 SOFS-R and administrative data estimate of earnings changes is explained by underreporting, but this is because the baseline discrepancy in estimates is so much larger.

Analysis of Civilian Earnings

As noted above, aligning the civilian earnings definitions in the SOFS-R and administrative data was complicated by the fact that SSA earnings are reported annually. For pre-activation earnings, we compared the SOFS-R estimates of civilian earnings to average monthly earnings received in the year before the activation as recorded in the administrative data. For the 2004 SOFS-R, the estimate of civilian earnings before activation in the survey was $890 (29 percent) higher than in the administrative data.

We could compute a comparable estimate of civilian earnings received during the activation period only for reservists whose activation spanned a full calendar year. In this limited sample, we found that average monthly civilian earnings during activation in the administrative data were $264 (34 percent) higher than in the 2004 SOFS-R.

These differences might reflect misreporting in the SOFS-R, but the difficulty in aligning the civilian earnings definitions makes it difficult to draw this conclusion with total confidence. In addition, the possibility that SOFS-R respondents are reporting pre-activation income not captured in SSA earnings records also prevents us from confidently attributing these civilian earnings differences solely to misreporting in the SOFS-R.

Comparison of 2004 and 2005 SOFS-R Earnings Estimates

Estimated earnings losses are much larger in the 2005 SOFS-R than in the 2004 SOFS-R. Our research suggests that this difference between the two waves of the SOFS-R is due to a few respondents who reported very large pre-activation earnings in the 2005 SOFS-R. The earnings questions in the 2005 SOFS-R asked respondents to report average earnings in the 12 months before activation whereas the 2004 SOFS-R did not specify the period over which average pre-activation earnings were to be computed. We conjecture that this change in question wording resulted in some respondents mistakenly reporting annual totals instead of monthly averages. A simple adjustment to the 2005 SOFS-R earnings data (dividing values that appear to be annual figures by 12) produces a distribution of earnings that closely resembles the earnings distribution in the 2004 SOFS-R.

Nonresponse Bias

The response rate to the 2004 and 2004 SOFS-R was 34 and 30 percent, respectively, which raises the possibility that the SOFS-R contains a select sample of reservists whose earnings experiences do not generalize to the full population of reservists. Our analyses in fact indicate that survey nonrespondents are quite different from survey respondents. Unweighted comparisons indicate that SOFS-R respondents are more likely than SOFS-R nonrespondents to be officers and in more senior pay grades and that average earnings as computed in the administrative data are 20 to 40 percent higher among SOFS-R respondents than nonrespondents. However, this differential nonresponse explains little of the difference between earnings change estimates in the SOFS-R and administrative data. This is because the influence of nonresponse bias is "differenced out" when computing earnings changes. Moreover, when SOFS-R survey weights are applied, the difference in mean earnings levels between survey respondents and nonrespondents diminishes substantially. The effectiveness of the SOFS-R survey weights further reduces the substantive importance of nonresponse bias in explaining differences between the two sets of earnings change estimates.

Implications

The empirical findings reported here have a number of implications. First, analysts and policymakers should employ SOFS-R data on military earnings with caution, in part because the SOFS-R earnings data do not include the value of the federal tax advantage. This issue becomes especially important when analyzing earnings during activation, since many of the pays and allowances reservists received while activated are tax exempt. A second reason is that SOFS-R respondents appear to significantly underreport military earnings. The omission of the tax advantage and underreporting of military earnings help explain why the SOFS-R data imply average earnings losses rather than the average earnings gains implied by the administrative data. Our analyses do not permit us to determine whether the SOFS-R respondents also misreport civilian earnings.

For these and other reasons, we believe that military personnel analysts should employ administrative data when feasible. Processing pre-existing administrative data is less expensive and less time-consuming than collecting comparable survey data. Furthermore, administrative data on earnings are likely to be more accurate than self-reported earnings recorded in surveys, although analysts should also be aware that administrative data can miss some sources of earnings (for example, under-the-table earnings). A significant limitation of administrative data is the relatively small amount of information it contains about the study population. Certain critical objective characteristics of the study population may not be contained in available administrative data sources. And subjective data, such as reenlistment intentions, can be collected only by survey. Thus, the best option available to the analyst may often be to match administrative data on key objective characteristics to survey data containing a richer array of respondent characteristics, intentions, and attitudes.

Finally, our results have methodological implications for survey data collection. We find that although response rates are low, the SOFS-R survey weights are able to correct for much of the resulting nonresponse bias in mean earnings. Consequently, it may be advisable for DMDC to devote more effort to minimizing the misreporting of survey items than to improving survey and item response rates. For example, if earnings questions are included, it could be advisable to ask separate questions about separate sources of earnings. This conclusion regarding nonresponse bias may not generalize to surveys of other populations, in part because weighting characteristics that are strongly related to earnings (such as pay grade) are not typically known for entire sample populations.

Acknowledgments

This research would not have been possible without the assistance of dedicated staff at the Defense Manpower Data Center (DMDC), the Social Security Administration (SSA), and Reserve Affairs within the Office of the Secretary of Defense (OSD-RA). We are grateful to Timothy Elig, Brian Lappin, and Sally Bird at DMDC who assisted us in preparing a data-protection plan for the project, shepherding our request to match survey and administrative data through DMDC's Human Subjects Protection Committee, implementing the match for us, and facilitating data transfer to SSA. We are indebted to Michael Risha at SSA for his continuing support of our research on the earnings of reservists. At OSD-RA, John Winkler, Tom Bush, and Col. Nilda Urrutia provided invaluable guidance throughout the course of the project.

Craig Martin at RAND oversaw data management for the project including obtaining and processing military personnel records, writing analysis programs, and facilitating data transfer to and from DMDC and SSA. His assistance on this and other projects related to the earnings of reservists has been instrumental and we thank him profusely for his patience and commitment to this research.

Abbreviations

ADP	automated data processing
ADPF	Active Duty Pay File
CPS	Current Population Survey
DMDC	Defense Manpower Data Center
DoD	Department of Defense
GWOT	Global War on Terror
LKM	Loughran, Klerman, and Martin
MEF	Master Earnings File
NDAA	National Defense Authorization Act
OIF	Operation Iraqi Freedom
ONE	Operation Noble Eagle
OSD	Office of the Secretary of Defense
OSD/P&R	Office of the Secretary of Defense–Personnel and Readiness
OSD/RA	Office of the Secretary of Defense–Reserve Affairs
RA	Reserve Affairs
RMC	Regular Military Compensation
RPF	Reserve Pay File
SOF	Status of Forces
SOFS-R	Status of Forces Survey of Reserve Component Members
SSA	Social Security Administration
SSN	Social Security Number
TPU	Troop Program Units
UI	Unemployment insurance
WEX	Work Experience File

CHAPTER ONE

Introduction

The reserve forces have been employed extensively during the Global War on Terror (GWOT). Large numbers of reservists have been called to active duty and the average duration of these active-duty spells has been long by historical standards (Loughran, Klerman, and Savych, 2005). Reservists experience a variety of hardships while activated, among which is the possibility that their labor market earnings might fall while they are activated.[1]

Administrative and survey-based data sources generate contradictory results regarding the effect of activation on reserve earnings (Loughran, Klerman, and Martin, 2006). The 2004 Status of Forces Survey of Reserve Component Members (SOFS-R) implies that about half of all activated reservists experience an earnings loss while they are activated and for most of those reservists, the earnings loss is large (more than 10 percent of their earnings before activation). In contrast, administrative data (combining Social Security Administration (SSA) earnings data with military pay data) suggest that most reservists experience large earnings gains and that earnings losses are relatively rare.[2]

In this report, we attempt to reconcile estimates of how the earnings of reservists change when they are activated as derived from SOFS-R data with analogous estimates derived from administrative data.[3] To do so, we match survey responses from the 2004 and 2005 SOFS-R to the type of administrative data on civilian and military earnings employed by Loughran, Klerman, and Martin (2006)—hereafter referred to as "LKM"—which allows us to directly compare estimates of earnings changes across the two data sources.

1 We use the term "activated" throughout this report to refer generically to a state of serving on active duty as a reservist in support of the GWOT and its specific contingencies (Operation Noble Eagle, Operation Enduring Freedom, and Operation Iraqi Freedom). An activated reservist may or may not be deployed. Being deployed generally means serving outside the continental United States in support of a specific contingency. In most cases, deployed also means serving in an officially designated combat zone.

2 See Appendix A for estimates of earnings changes attributable to activation derived from administrative data by year activated, activation duration, pay grade, and military occupation. There are numerous examples where administrative and survey data generate conflicting empirical results. For instance, Shochet, McConnell, and Burghardt (2003) find that the positive program effects of the Job Corps program found in survey data are not found in administrative earnings records. Other recent research documenting substantive discrepancies between survey and administrative data include Goldman and Smith (2001), Denmead and Turek (2005), Hurd and Rohwedder (2006), Kapteyn and Ypma (2007), and Haider and Loughran (2008).

3 None of the estimates reported in the main text of this report should be interpreted as estimates of the causal effect of activation on the earnings of reservists. Instead, they should be interpreted as descriptive estimates of how, on average, reserve earnings change between the periods before activation and during activation. Causal estimates require an estimate of counterfactual changes in earnings, which cannot be generated employing SOFS-R data, since the SOFS-R asks earnings questions only of reservists who are activated (see Chapter Two). See LKM and Appendix A for causal estimates of the effect of activation on earnings.

At first glance, it might seem that administrative data are more likely than survey data to produce accurate estimates of earnings change. The administrative earnings data we employ records earnings as reported directly by the Department of Defense (DoD) and civilian employers. Moreover, these employer reports are typically generated from the same computerized systems used to generate paychecks. By contrast, SOFS-R earnings are reported by reservists themselves and reservists may misreport earnings for any number of reasons (e.g., systematic omissions, misunderstanding the question language). In addition, estimates derived from the SOFS-R are potentially subject to systematic survey and item nonresponse bias, a potential problem in all surveys. However, it is important to recognize that administrative data are not perfect either. For example, our administrative data do not include earnings not reported to SSA, such as unreported tips or other under-the-table earnings.

The remainder of this report is organized as follows. Chapter Two describes how we construct our matched data file. Chapter Three then quantifies the degree to which differences in the treatment of the tax advantage and misreporting of military and civilian earnings in the SOFS-R explain observed differences in estimates of earnings changes. Chapter Four contains a separate analysis of nonresponse bias in the SOFS-R and Chapter Five presents conclusions.

CHAPTER TWO

Data and Methods

This chapter describes the SOFS-R first and then our administrative data. Having described the two data sources, the chapter then discusses how we merge them together to create our analysis file.

SOFS-R

The Status of Forces Surveys, administered by the Defense Manpower Data Center (DMDC), are a suite of periodic surveys of active and reserve component members and DoD civilian employees. They are designed to track the opinions, attitudes, and experiences of DoD military and civilian personnel. The SOFS-R is conducted online and is designed to be representative of individuals actively serving in the Selected Reserves.[1]

This study employs the May 2004 and June 2005 SOFS-R. Those surveys included questions concerning periods of active-duty service and earnings before and during active-duty service. The 2004 SOFS-R asks whether respondents had been activated in the 24 months before the survey (including activations that began more than 24 months before the survey), and the 2005 SOFS-R asks about activations after September 11, 2001.[2] Reservists who had been activated were then asked a series of questions about their labor market earnings. Specifically, they were asked to report their average monthly civilian and military earnings before, during, and after their most recent activation (a total of six questions).[3] About 20 percent of respondents reported pre-activation civilian earnings by providing a range rather than a specific number, and just under 30 percent answered the questions on military earnings with a range. Overall, about 40 percent of respondents answered at least one of the earnings questions by providing

[1] Reservists who had less than six months of service when the survey was conducted or who were of flag rank when the sample was drawn (six months before the survey) were excluded from the survey. Reservists who were selected to participate in the survey were notified by mail one month before the survey was actually administered and second notifications were issued via email within 24 hours after the questionnaire was posted on the website. Sampled individuals who did not return a completed survey were sent up to six reminder emails and three reminder letters. For more information about the SOFS-R, please refer to Defense Manpower Data Center (2004, 2005).

[2] The 2005 SOFS-R also includes the month in which the most recent activation began and, if it ended, the month in which it ended.

[3] Respondents are instructed to report their average monthly civilian "income" and average monthly military "compensation."

a range. When respondents did not provide an actual dollar amount, we used the midpoint of the reported range.[4]

The 2004 and 2005 SOFS-R differ in several important ways. First, nearly four times as many reservists were sampled for the 2005 SOFS-R (211,003) than for the 2004 SOFS-R (55,794).[5] Second, the 2005 SOFS-R asks about activations after September 11, 2001, whereas the 2004 SOFS-R asks only about activations in the preceding 24 months. To focus on comparable samples of activated reservists, we limit our 2005 SOFS-R sample to those reservists who were activated in the preceding 24 months.[6]

Third, the wording of the earnings questions differs across the two years. In particular, when asking about earnings before activation, the 2005 SOFS-R instructs respondents to report average monthly income in the 12 months before the most recent activation, but the 2004 SOFS-R does not specify a time period.[7] The wording of the questions about earnings during activation remained largely unchanged between the two surveys. As the evidence presented in Chapter Three suggests, this change in the wording of the questions about earnings before activation appears to have sharply increased estimates of the fraction of reservists reporting earnings losses between the 2004 and 2005 SOFS-R.

Both surveys have relatively low response rates. The SOFS-R's unweighted response rate (i.e., the fraction of eligible surveyed reservists who responded to the survey and answered the question about whether they had been activated in the preceding 24 months) is 34 percent in the 2004 SOFS-R and 30 percent in the 2005 SOFS-R.[8] Among activated reservists who responded to the survey, about one-fifth do not have valid answers for all of the earnings questions (19 percent in the 2004 SOFS-R and 23 percent in the 2005 SOFS-R). In Chapter Four, we examine the substantive importance that any bias survey and item nonresponse may impart to the SOFS-R estimates of earnings changes attributable to activation.

The bulk of the analyses reported here use data on SOFS-R respondents who gave valid answers to all four questions on earnings received before and during activation. There are 9,514 such respondents to the 2004 SOFS-R and 37,310 respondents to the 2005 SOFS-R. Below, we discuss additional sample restrictions arising from an inability to match survey and administrative records and because of difficulties matching periods of active-duty service defined in the two data sources.

---

4 In both the 2004 and 2005 SOFS-R, the median range was $500 for military earnings during activation and civilian earnings before activation, $150 for military earnings before activation, and $400 for civilian earnings during activation.

5 This increase in sample size was made in part because the 2005 National Defense Authorization Act (NDAA) required that DoD conduct a survey of at least 50 percent of Selected Reservists. One objective of the 2005 SOFS-R was to provide data that could be used to study the effect of activation on the earnings of reservists.

6 According to self-reported information on the starting month and duration of the most recent activation, 3.6 percent of respondents who were activated after September 11, 2001, and who were eligible to answer the questions on earnings would not be included in our analysis because the activation ended more than 24 months before the survey was conducted.

7 The wording change was in response to a legislative mandate (contained in the 2005 National Defense Authorization Act) to study the change in earnings that occurs during activation relative to average earnings in the 12 months before activation. The 12-month pre-activation reference period was specifically stated in the legislation. See Appendix B for the wording of all the earnings questions used in this study.

8 Weighted response rates were 39 percent in the 2004 SOFS-R and 42 percent in the 2005 SOFS-R (Defense Manpower Data Center, 2004, 2005).

Administrative Data

The dataset we construct from administrative data sources links reserve personnel records to information on activations and earnings. To identify samples of reservists, we use DMDC's Work Experience File (WEX). The WEX is generated from DMDC's Active Duty Military Personnel Master File and Reserve Component Common Personnel Data System File and contains at least one record for every individual serving in the active or reserve components on or after September 30, 1990.[9] From this file, we determine enlistment status, pay grade, unit, and component in each month.

Information on activations and deployments comes from DMDC's GWOT Contingency File (henceforth, "Contingency File"). The Contingency File is intended to include a record for every activation or deployment after September 11, 2001, in support of the GWOT. Each record in the file includes the start and end date of each activation or deployment. Generally, deployments are nested within an activation spell. However, some deployments occur without a corresponding activation spell or are not nested within an activation spell.[10] In these cases, we use the union of activation and deployment spells even though the survey questions on earnings reference only activation spells. We took this approach for two reasons. First, the text of the survey questionnaire at the beginning of the section containing the earnings questions indicated that the information being collected would be used to "better assess the financial impact of activation/deployment on members" (underlining in original). Second, we did not want to miss any activations that were miscoded in the Contingency File as deployments.

A drawback to using the Contingency File to define activation spells is that it includes information only on activations in support of the GWOT. Thus, survey respondents who were activated for other contingencies during this time period (e.g., operations in Bosnia) will not appear as being activated in the administrative data. An alternative to using the Contingency File is to use pay data to infer periods of activation. However, we found that it was difficult to identify activation spells reliably with the pay data. The pay data frequently generate very short activation spells when the Contingency File and the 2005 SOFS-R data indicate much longer activation spells. As we discuss below, correctly identifying the timing and length of activations is essential for aligning the survey and administrative data earnings definitions. Therefore, we decided to use the Contingency File as our source of information on activations. Even though the Contingency File misses activations that were not in support of the GWOT, the estimates we report here of the change in earnings during activation are similar in magnitude to estimates reported in LKM, which cover all activations.[11]

To measure military earnings, we link the personnel records to the Reserve Pay File (RPF) and the Active Duty Pay File (ADPF).[12] These files include information on all military

---

[9] The file contains military personnel transaction records back through 1975.

[10] Six percent of reservists in the Contingency File had a record indicating that they were deployed without a corresponding activation record. This might happen for brief deployments that occur near a reservist's residence and do not involve a call-up to active duty.

[11] LKM report that annual earnings increase between 2000 and 2003 by an average of $15,647, or $1,303 per month, for reservists activated in 2003. Below, we report that, according to administrative data, reservists in the 2004 SOFS-R experience average monthly earnings gains of $1,379 per month in the year they are activated relative to the year immediately preceding activation.

[12] The ADPF contains the military earnings of activated Navy and Marine Corps reservists and the RPF contains the military earnings of all other reservists.

pays, bonuses, and military allowances.[13] Our measure of total military earnings is obtained by summing over all military pays and allowances but excluding bonus payments.[14] Bonuses are excluded because the 2005 SOFS-R instructs respondents to report earnings net of bonus payments. Although the 2004 SOFS-R does not have any explicit directions regarding bonuses, we exclude bonuses from calculations involving the 2004 SOFS-R to facilitate making comparisons across the two surveys.[15]

Data on civilian earnings come from SSA's Master Earnings File (MEF). The MEF contains information reported to SSA by employers on earnings subject to Medicare taxes. Almost all earnings are subject to Medicare taxes, so this database has nearly universal coverage of all civilian employment in the United States. In the administrative data, we compute civilian earnings as earnings recorded in the MEF minus military earnings subject to Medicare taxes (where military earnings subject to Medicare taxes are recorded in the RPF and the ADPF).[16] We then linked records from the various administrative data files using scrambled Social Security Numbers (SSNs).[17]

The primary advantage of administrative earnings data lies in their quality. The RPF and ADPF are the files used to generate military paychecks. Therefore, they record military earnings actually received by military personnel. The MEF data are reported to SSA by employers. In most cases, these reports are generated by the same computer systems that generate civilian paychecks. Incorrect reporting is subject to civil and criminal penalties. Thus, it seems reasonable to assume that the MEF records earnings values that are quite close to earnings actually received.[18] In contrast, there is some evidence that survey data on earnings diverge systematically from payroll records.[19]

The MEF data, however, have two important limitations. First, they are available only on a calendar year basis. The SOFS-R, in contrast, asks for average monthly civilian earnings over a period that does not necessarily correspond to a particular calendar year. As we discuss below, this limitation makes it difficult to align the civilian (and therefore total) earnings definitions in the SOFS-R and administrative data. However, we do have information on monthly military earnings, which allows us to align the military earnings definitions quite closely.

---

13 Military pays include basic pay, drill pay, and hostile fire/imminent danger pay. Allowances include basic allowance for housing, basic allowance for subsistence, and family separation allowance.

14 We treat one dollar of income from basic pay the same as one dollar of allowance or special pay income. However, points toward the military's pension system do not accrue for allowance income. Although the pension implications might therefore differ across types of military compensation, this report focuses on earnings so this is not an issue for our analysis.

15 The discrepancy between estimates generated using the SOFS-R and administrative data is slightly larger when bonuses are included in the administrative data.

16 All military pays other than allowances are reported to SSA.

17 DMDC did not provide RAND with actual SSNs to protect the privacy of military personnel.

18 This point is made by researchers who have used administrative earnings records to assess the validity of survey data (Bound and Krueger, 1991; Baj, Trott, and Stevens, 1991; Hill et al., 1999).

19 Roemer (2000) compares self-reported wage income in the Current Population Survey (CPS) to tax returns submitted to the Internal Revenue Service and finds that the survey data are reasonably accurate in the middle of the income distribution but that the survey and tax return records differ substantially among higher- and lower-income respondents. Bound and Krueger (1991) compare CPS self-reports to SSA earnings and find close to zero net bias. Rodgers, Brown, and Duncan (1993) examine the earnings of respondents to the Panel Study of Income Dynamics who were unionized employees of a single firm and find that "usual" and weekly earnings are systematically misreported. Annual earnings were reported with less error. See Hotz and Scholz (2002) for a summary of this line of research.

Second, the MEF might not capture all civilian earnings. As just discussed, almost all civilian labor market earnings should be reported to SSA and hence recorded in the MEF. However, some reservists might receive income under the table that is easily concealed and so might not be recorded in the MEF (Hotz and Scholz, 2002). In Chapter Three, we discuss the existing research on this issue and its substantive importance for our research.

Matching the SOFS-R and Administrative Data

To safeguard the privacy of SOFS-R respondents, DMDC data-protection procedures prohibited RAND from gaining direct access to the matched survey and military personnel data. In addition, SSA never releases individual earnings data. Therefore, we followed a multistep procedure to build our matched file and analyze those data.

First, we processed the military personnel records obtained from DMDC at RAND. Second, we sent those data to DMDC and their analysts matched our processed military personnel record files to the SOFS-R sample file employing a scrambled SSN. Match rates (reported in Table 2.1), were high but not perfect. For respondents who answered all of the relevant earnings questions, DMDC found matching records in the RAND military personnel files for 95.2 percent of the 2004 SOFS-R respondents and 98.2 percent of the 2005 SOFS-R respondents. Weighted match rates were 98.1 percent and 97.4 percent, respectively. For reservists who did not answer the earnings questions, we matched 94.2 percent of the 2004 SOFS-R nonrespondent sample and 97.7 percent of the 2005 SOFS-R nonrespondent sample. The nonrespondent sample includes those who did not respond to the survey at all and those who did not respond to all of the relevant survey questions (either because they did not answer the earnings questions or were not asked the earnings questions because they were not activated).

DMDC then sent the matched SOFS-R and military personnel data to SSA. SSA matched the file received from DMDC to individual annual earnings data reported in the MEF. SSA

Table 2.1
SOFS-R and DMDC Administrative Data Match Rates

	Earnings Items Respondents	Survey Nonrespondents, Not Activated Respondents, and Earnings Item Nonrespondents
	A. 2004	
Total number of survey records	9,514	44,629
Matched records	9,058 (95.2%) [98.1%]	42,034 (94.2%) —
	B. 2005	
Total number of survey records	37,310	167,447
Matched records	36,639 (98.2%) [97.4%]	163,447 (97.7%) —

NOTES: The unweighted percentages of matched records are shown in parentheses. For respondents, the weighted percentages of matched records are shown in brackets (survey weights are not available for nonrespondents).

has records only for individuals with earnings. All reservists who were activated should have received military pay and, therefore, should have had at least one SSA earnings record. We successfully matched over 99 percent of the matched SOFS-R and military personnel data records to MEF earnings records for both the 2004 and 2005 survey waves.[20]

Finally, SSA executed programs supplied by RAND that analyzed the completed matched analysis file. Those programs generated group-level statistical output (e.g., means and variances) that were then returned to RAND for further analysis. Output was generated for groups defined by rank, survey response status (sampled respondent, sampled nonrespondent, not sampled), whether they had been activated in the year before their most recent activation, and whether their most recent activation spanned an entire calendar year (see the discussion in Chapter Three for the reasoning behind these later groupings).

Analysis of Differences in Estimated Earnings Change During Activation

The balance of this report relates the results of analyzing the matched SOFS-R and administrative data in an effort to understand the divergence between SOFS-R and administrative estimates of earnings changes.[21] We attribute differences between the two sets of earnings change estimates to two sources: (1) SOFS-R respondents may not report earnings accurately and (2) SOFS-R respondents may differ from those who were in the SOFS-R sample but did not respond to the survey or the earnings questions. Chapter Three considers the first source of discrepancy and Chapter Four considers the second.

---

[20] A scrambled SSN is used to carry out the merge with the MEF, and this SSN is missing whenever an SOFS-R record was not matched to the DMDC personnel data.

[21] Unless otherwise noted, all estimates employ the SOFS-R survey weights that adjust for design effects (i.e., oversampling of certain subgroups) as well as differential nonresponse.

CHAPTER THREE

Decomposing Differences in Estimated Earnings Changes

In this chapter, we decompose the difference in SOFS-R and administrative estimates of reservists' earnings changes while they are activated into those attributable to three sources: differential treatment of the tax advantage, misreporting of military earnings in the SOFS-R, and misreporting of civilian earnings in the SOFS-R. Differences attributable to survey nonresponse are addressed in Chapter Four. We begin by establishing a baseline estimate of the difference in earnings changes employing the sample of SOFS-R respondents matched to military and civilian administrative earnings data. We then quantify the degree to which the three sources just listed account for differences in the baseline estimates of earnings changes.

Baseline Difference in Estimates of Earnings Changes

We begin by computing a baseline estimate of the change in earnings between the periods before and during activation employing the SOFS-R data and then employing the administrative data for the same individuals. Table 3.1 reports the results of these computations.[1]

Table 3.1
Estimates of Average Monthly Earnings Change Derived from SOFS-R and Administrative Data (in dollars)

	SOFS-R	Administrative	SOFS-R Minus Administrative	% Difference
		A. 2004		
Total before activation	3,714	3,182	532	17
Total during activation	3,428	4,561	–1,133	–25
Earnings change	–287	1,379	–1,666	–121
		B. 2005		
Total before activation	12,086	3,000	9,086	303
Total during activation	6,463	4,626	1,878	40
Earnings change	–5,623	1,625	–7,248	–446

NOTE: Percentages are calculated relative to administrative data totals.

[1] Ideally, all calculations would be in real dollars. However, we do not know the year in which earnings were received for the survey, so we cannot determine the appropriate deflator. Therefore, all calculations are in nominal dollars. This is unlikely to pose a problem, as the report focuses on a comparison of the survey and administrative data results and both are in nominal dollars.

First, consider the estimates derived from SOFS-R data. The SOFS-R questions directly ask about average monthly military and civilian earnings before and during activation. Weighted tabulations show large declines in average monthly earnings between those two periods: $287 in the 2004 SOFS-R and $5,623 in the 2005 SOFS-R. The large negative earnings change estimate derived from the 2005 SOFS-R data is due to the implausibly large estimate of monthly earnings before activation of $12,086. This estimate corresponds to average annual pre-activation earnings of more than $145,000, which is much too high for this population.

To generate a comparable estimate of earnings changes employing the administrative data, we needed to identify the period of activation to which the SOFS-R refers. However, we know only that in both the 2004 and 2005 SOFS-R, the relevant activation period occurred in the 24 months before the survey date (May 2004 and June 2005).[2] So, for each SOFS-R respondent, we first identified the most recent activation period in the Contingency File in the 24 months before the survey dates. We were able to identify an activation period in the Contingency File for 76 percent (85 percent unweighted) of 2004 SOFS-R respondents and 82 percent (91 percent unweighted) of 2005 SOFS-R respondents.[3]

Next, we needed to choose reference periods over which to compute average monthly earnings before and during activation in the administrative data. Because only annual civilian earnings data are available, we generally do not observe administrative earnings for periods that coincide exclusively with the periods before and during activation. Instead, we treated the calendar year during which the majority of the activation took place as the during activation year and the year before the year in which the activation began as the "pre" or before-activation year. For instance, for a reservist activated in August 2003 through March 2004, the pre-activation year would be 2002 and the during-activation year would be 2003. Clearly, the administrative estimate of earnings during activation will be biased up or down depending on whether those earnings are in fact lower or higher than earnings before activation. We discuss this issue further below.

Having identified the appropriate reference periods, we define total earnings in the administrative data as the sum of civilian earnings, military earnings (pays and allowances), and any tax advantage of military compensation (see immediately below for an explanation). The earnings change is then computed by taking the difference between total earnings received in the during-activation year and earnings in the year before activation (converted to a monthly figure by dividing by 12).

Given these definitions, the estimates based on the administrative data indicate large average earnings gains during activation. For the 2004 matched sample, average monthly earnings increase $1,379, an increase of 40 percent over earnings before activation, whereas for

---

[2] The 2005 SOFS-R identifies the starting month of the activation as well as its duration (in months). This information is not available in the 2004 SOFS-R.

[3] Survey misreporting is one reason why activation status differs in the two data sources. For instance, SOFS-R respondents may have included activations that ended more than 24 months before the survey date (the type of response error where respondents report events as having occurred more recently than they actually did is known as "telescoping"; see Bound, Brown, and Mathiowetz, 2001, p. 3744). It also could be that some activations in support of the Global War on Terror are erroneously not recorded on the the Contingency File. Finally, respondents might be referring to an activation that was in fact not in support of the GWOT and would not be covered by the Contingency File. Reservists who claimed to have been activated, but for whom we did not locate an activation record in the Contingency File, have mean self-reported pre-activation earnings in the 2004 SOFS-R that are about 5 percent lower than they are for reservists who were activated according to the Contingency File. Earnings during activation are 14 percent lower in this sample.

the 2005 matched sample, earnings increase by $1,625, an increase of more than 50 percent over earnings before activation. These average earnings gains are broadly consistent with those reported in LKM. Note that unlike estimates based on the SOFS-R data, these administrative data estimates of earnings changes are reasonably similar across the two years.

Comparing the two sets of estimates, we see that the 2004 SOFS-R implies monthly earnings changes that are $1,656 smaller than those implied by administrative data. The 2005 SOFS-R implies monthly earnings changes that are $7,248 smaller than those implied by the administrative data.

The remainder of this chapter attempts to account for the baseline differences in estimates of earnings changes reported in Table 3.1. Our analysis assumes that the earnings definitions employed in the two data sources are identical, but the preceding discussion clearly suggests otherwise. First, the numbers based on administrative data include a measure of tax advantage. Although tax advantage is not typically considered a component of earnings, the baseline estimates do include it, since the tax advantage is considered part of Regular Military Compensation (RMC). In contrast, the survey estimates refer to pretax amounts.

Second, the administrative data correspond to earnings received in the year of activation, whereas the SOFS-R data correspond to earnings received during the actual period of activation. LKM show that earnings gains increase with days of active-duty service in a given year. It is likely, then, that estimates based on administrative data will tend to understate earnings gains for the matched SOFS-R sample. Thus, if we were able to measure earnings during activation perfectly using the administrative data, the difference in estimates of earnings changes derived from SOFS-R and administrative data would likely be even larger than what we report in Table 3.1. This further suggests that we would be able to account for less of the absolute difference in estimates than our analyses in the following sections imply (although it is not clear whether our adjustments would explain a larger or smaller *percentage* of the discrepancy).

Difference in Estimates of Earnings Changes Attributable to the Tax Advantage

Military allowances and all military pays received while serving in a combat zone are not subject to federal income taxes. Following the definition of RMC used in the federal "Green Book," we define the tax advantage as the amount of additional income one would need to receive to make after-tax income without the preferential tax treatment equal to what after-tax income would be with the preferential tax treatment.[4] Estimates of earnings changes employing the administrative data account for the value of this tax advantage under the assumption that reservists file as single with no dependents.[5]

The SOFS-R, however, instructs respondents to report pretax earnings. This is potentially an important limitation for the survey data. The tax advantage is a component of Regular

[4] See Office of the Under Secretary of Defense (Comptroller) (2005).

[5] See LKM for additional details on how this calculation was implemented. The assumption that reservists file as single with no dependents is clearly not valid, but we lack the data on marital status and spousal earnings needed to relax this assumption. However, the effect of this assumption on our tax imputations is likely to be small, on average. On one hand, assuming that reservists are unmarried means that spousal earnings do not affect the reservists' marginal tax bracket. All else equal, this assumption lowers estimated taxes. On the other hand, the assumption that reservists have no dependents reduces the number of exemptions the reservist can declare, which raises estimated taxes.

Military Compensation and, since earnings received before and during activation are treated differently for tax purposes, changes in pretax earnings could provide a misleading assessment of the change in take-home pay.

Ideally, we would estimate any tax advantage based on the survey information and add it to the survey estimates of earnings change. However, doing so is difficult. Calculating the tax advantage requires knowing, at a minimum, total calendar year income broken out by whether it is subject to federal income taxes. The survey does not distinguish between taxable and non-taxable earnings, and the earnings variables refer not to calendar years but to the period before and during the most recent activation.

Thus, to estimate the quantitative importance of the tax advantage in explaining differences between the SOFS-R and administrative estimates of earnings changes, we examine how the baseline difference between the survey and administrative estimates changes when we exclude the tax advantage from our computation of total earnings in the administrative data.

The advantage of this approach is that it cleanly identifies the quantitative significance of the different way tax advantage is treated in the survey and in the administrative data calculations. The disadvantage, as noted above, is that the change in gross earnings ignores the differential tax treatment and overstates earnings losses (understates earnings gains), since a greater share of earnings during activation receives the preferential tax treatment. However, the point of this exercise is to understand why the survey and administrative data produce such different answers. Any estimate of tax advantage from the survey would be problematic and not exactly comparable to the estimate from the administrative data. Adding a flawed estimate of the tax advantage to the survey estimate would then introduce another source of discrepancy between the administrative and survey estimates of earnings changes.

Table 3.2 recomputes differences in earnings changes excluding the tax advantage from the administrative data. This adjustment to the administrative data closes the gap between the administrative and survey estimates of earnings changes by nearly $500 in the 2004 sample and by nearly $600 in the 2005 sample. Thus, accounting for differences in the treatment of the tax advantage reduces the discrepancy between the baseline SOFS-R and administrative

Table 3.2
Estimates of Average Monthly Earnings Change Derived from SOFS-R and Administrative Data, Excluding the Tax Advantage (in dollars)

	SOFS-R	Administrative Data	SOFS-R Minus Administrative Data	% Difference
		A. 2004		
Total before activation	3,714	3,095	619	20
Total during activation	3,428	4,005	–577	–14
Earnings change	–287	910	–1,196	–132
		B. 2005		
Total before activation	12,086	2,924	9,162	313
Total during activation	6,463	3,986	2,478	62
Earnings change	-5,623	1,062	–6,684	–630

NOTE: Percentages are calculated relative to administrative data totals.

estimates by about 28 percent in the 2004 sample and by about 8 percent in the 2005 sample. Note that in terms of dollars, this adjustment has similar effects across the two years.

Difference in Estimates of Earnings Changes Attributable to Misreported Military Earnings

In this section, we directly compare estimates of military earnings derived from the SOFS-R and administrative data for the same individuals. We attribute any difference in these estimates of military earnings to misreporting in the SOFS-R under the assumption that the administrative data record military earnings accurately and that we can perfectly align the military earnings definitions in the two data sources. In the sections below, we describe how we align the military earnings definitions, present estimates of the quantitative importance of misreporting in the SOFS-R, and argue how changes in question language led to the large differences in estimates of military earnings recorded in the 2004 and 2005 SOFS-R.

Aligning the Military Earnings Concepts

To align the SOFS-R and administrative military earnings definitions, we must define the period to which the SOFS-R responses on military earnings refer. For the period during activation, we simply take the average monthly military earnings for the time that the Contingency File indicates a given reservist was activated. In some cases, the administrative data indicated that a reservist had no military earnings in months in which, according to the Contingency File, he or she was serving on active duty. Since all reservists on active duty should have some military earnings, we also computed monthly military earnings while activated in the administrative data using only data for months in which the reservist had positive military earnings.

Aligning the pre-activation period is more complicated. The 2004 SOFS-R does not provide respondents with a specific pre-activation period over which to report military earnings. Instead, the 2004 SOFS-R instructs respondents to report average earnings before the most recent activation. A natural reading of this wording would appear to imply that this average should not include periods from an earlier activation. The 2005 SOFS-R requested that respondents report average monthly military earnings for the 12 months before the most recent activation. Therefore, for both surveys, we take as our primary estimate of pre-activation earnings average monthly earnings in the 12 months before the activation, excluding any other activation during that period. We also report estimates using two other reference periods: the month immediately before activation and the 12 months before activation including any previous activation.

The Quantitative Importance of Misreported Military Earnings in the SOFS-R

Table 3.3 reports average monthly military earnings as computed using SOFS-R and administrative data for the same individuals. In the 2004 sample (Panel A), average monthly military earnings are about 15 percent higher in the administrative data when we average military earnings over the 12 months before the activation and exclude any months during an earlier activation. During activation, military earnings are 20 percent higher in the administrative data than in the SOFS-R data when we exclude activation months during which reservists had zero earnings.

Table 3.3
Comparison of Mean Military Earnings in SOFS-R and Administrative Data (in dollars)

	SOFS-R	Administrative Data	SOFS-R Minus Administrative Data	% Difference
A. 2004, Earnings Before Activation				
Month before activation	636	939	–303	–32
12 months before activation	636	910	–274	–30
12 months before activation, excluding months of earlier activation	627	734	–108	–15
B. 2004, Earnings During Activation				
All months	3,086	3,616	–531	–15
Excluding months with no military earnings	3,101	3,889	–788	–20
C. 2005, Earnings Before Activation				
Month before activation	972	4,003	3,031	312
12 months before activation	866	4,003	3,137	362
12 months before activation, excluding months of earlier activation	732	4,016	3,284	449
D. 2005, Earnings During Activation				
All months	5,977	3,569	2,408	67
Excluding months with no military earnings	5,794	3,827	1,967	312

NOTE: Percentages are calculated relative to administrative data totals.

Appendix C reports more detailed comparisons between the SOFS-R and administrative data estimates of military earnings. Those comparisons reveal that the difference in military earnings estimates cannot be explained by simple omissions in self-reports (e.g., reporting only basic pay). In addition, the analyses indicate that about two-thirds of 2004 SOFS-R respondents report military earnings that are less than those recorded in the administrative data.

Panels C and D of Table 3.3 compare military earnings reported in the 2005 SOFS-R and those recorded in the administrative data. There, we see, across all reference periods, that average monthly military earnings are much higher in the SOFS-R than in the administrative data. In the following section, we offer an explanation for why the 2004 and 2005 SOFS-R military earnings estimates differ so sharply.

The quantitative importance of misreporting military earnings in the 2004 SOFS-R is summarized in Panel A of Table 3.4. In that table, we compute earnings differences replacing the SOFS-R estimates of military earnings with the administrative estimates of military earnings. In the 2004 SOFS-R, adjusting military earnings for underreporting in the survey accounts for 14 to 42 percent of the baseline discrepancy in the mean earnings change estimates, with the 42 percent figure representing our preferred estimate (excluding prior activations and conditioning on positive military earnings while activated).

Our preferred estimates imply that the omission of the tax advantage and misreporting of military earnings in the 2004 SOFS-R together explain more than 70 percent of the difference

Table 3.4
Difference in Estimated Earnings Changes Accounted for by Tax Advantage and Misreported Military Earnings (in dollars)

Adjustment	SOFS-R Before Activation	SOFS-R During Activation	SOFS-R Difference	Administrative Minus SOFS-R	% Difference Explained by Adjustment
A. 2004					
None	3,714	3,428	–286.6	1,666	—
Remove tax advantage	3,714	3,428	–286.6	1,196	28
Replace SOFS-R military earnings with administrative data					
Month before	4,018	3,958	–59	969	14
Month before, conditional during	4,018	4,232	214	696	30
12 months before	3,989	3,959	–30	940	15
12 months before, conditional during	3,989	4,232	243	667	32
12 months before, excluding prior activation	3,813	3,958	146	764	26
12 months before, excluding prior activation, conditional during	3,813	4,232	419	491	42
B. 2005					
None	12,086	6,463	–5,623	7,248	—
Remove tax advantage	12,086	6,463	–5,623	6,684	8
Replace SOFS-R military earnings with administrative data					
Month before	9,055	4,056	–4,999	6,060	9
Month before, conditional during	9,055	4,314	–4,741	5,802	12
12 months before	8,949	4,056	–4,893	5,954	10
12 months before, conditional during	8,814	4,314	–4,501	5,562	15
12 months before, excluding prior activation	8,814	4,056	–4,759	5,820	12
12 months before, excluding prior activation, conditional during	8,814	4,314	–4,501	5,562	15

NOTES: The percentage difference explained is equal to the negative of the percentage difference between the baseline discrepancy and the discrepancy after the adjustment relative to the baseline discrepancy. "Conditional during" refers to average earnings in months during the activation where the administrative data indicate positive earnings.

in the baseline earnings change estimates. As noted in the previous section, though, the baseline difference between the SOFS-R and administrative estimates of earnings changes might be even larger if we could align the civilian earnings concepts perfectly. Thus, it is likely that the omission of the tax advantage and misreporting of military earnings in the 2004 SOFS-R explain less than 70 percent of the true baseline difference in estimates.

Explaining the Difference Between the 2004 and 2005 SOFS-R Estimates of Military Earnings

Although the 2004 SOFS-R estimates of military earnings lie below those recorded in administrative data, the 2005 SOFS-R estimates of military earnings are much higher than those recorded in the administrative data (see Panel B of Table 3.3). The 2005 SOFS-R estimates of military earnings in the period before activation are almost double the military earnings recorded in the administrative data ($3,000–$3,300 or 76–82 percent higher). The difference in military earnings during activation is also positive but smaller in magnitude ($2,000–$2,400 or 34–40 percent higher). Note that the administrative data indicate comparable levels of military earnings for both the 2004 and 2005 samples whereas the SOFS-R data suggest that military earnings grew sharply between the two years.

A closer examination of the distribution of the difference in military earnings estimated using the SOFS-R and administrative data reveals that in both years, most SOFS-R respondents underreport military earnings relative to what they report in the administrative data. Averaging military earnings over the 12 months before activation and excluding months from an earlier activation, the median difference in pre-activation military earnings as estimated by the 2005 SOFS-R and administrative data is $49. This difference is somewhat smaller than the same difference computed for the 2004 sample ($150). The median difference in during-activation military earnings is $227 in 2005, which is similar in magnitude to what was found for 2004 ($335).

These results indicate that the divergence between the 2004 and 2005 SOFS-R estimates is more pronounced for military earnings before activation than for earnings during activation and that this difference in results is likely due to outliers in the 2005 SOFS-R. Figure 3.1 plots the distributions of military earnings recorded in the 2004 and 2005 SOFS-R. The distributions nearly overlap in the lower range of military earnings before activation and then diverge in the upper range. Figure 3.1 also shows that the difference in these distributions is more pronounced in the case of earnings before activation than it is in the case of earnings during activation. The distributions of pre-activation military earnings begin to diverge at about the 60th percentile, and, for military earnings during activation, they begin to diverge around the 80th percentile.

Thus, an explanation for why the 2004 and 2005 SOFS-R have such different estimates of military earnings should also explain why the difference is larger for military earnings before activation than it is for military earnings after, as well as why the difference in distributions is evident only in the upper tails. A candidate explanation is based on a subtle change in the wording of the 2004 and 2005 SOFS-R questions that asked respondents about military earnings before their most recent activation.

The exact wording of the question in the 2004 SOFS-R was

> How much was your average <u>monthly military</u> compensation <u>prior</u> to your most recent <u>activation</u>, before taxes and other deductions? (underlining in original).

The corresponding question in 2005 was

> How much was your <u>average monthly military</u> compensation (excluding reenlistment bonuses) in the 12 months <u>prior</u> to your most recent <u>activation</u>, before taxes and other deductions (i.e., gross pay)? (underlining in original).

The primary change in the question wording is the addition of the phrase "in the 12 months" to the question in the 2005 SOFS-R. The wording of the questions about military earnings received during activation remained largely unchanged between the years.[6]

Although the intention of the wording change was to make the reference period more specific, it seems plausible that the change in wording caused some respondents to report annual (or 12-month) earnings totals rather than average monthly earnings over a 12-month period.[7] Thus, for example, a reservist with actual earnings in the 25th percentile of monthly earnings (i.e., $1,200) but who mistakenly reported annual earnings (i.e., $14,400) would appear to have military earnings in the 99th percentile of monthly earnings. If 20 to 25 percent of the observations in the 2005 SOFS-R are misreported in this way (below, we estimate that 21 percent of SOFS-R respondents misreported military earnings in this way), then the 2004 and 2005 distributions of military earnings would be reasonably close to each other except in the upper tail, and mean, but not median, military earnings would differ substantially across the two years. Moreover, since only the wording of the question about military earnings before activation changed, this would explain why the estimates differ more in the case of military earnings before activation than they do in the case of military earnings during activation.

Some of those who misreported military earnings before activation in this way may have made the same error when reporting military earnings during activation.[8] We cannot directly identify reservists who report annual totals in the 2005 SOFS-R, but we can employ the distribution of self-reported military earnings in the 2004 SOFS-R (which we assume is not afflicted by this type of reporting error) to identify outliers in the 2005 SOFS-R responses. Doing so, we find that 42 percent of 2005 SOFS-R respondents who reported pre-activation military earnings values above the 99th percentile of the 2004 distribution of pre-activation military earnings also reported during-activation military earnings that lay above the 99th percentile of the 2004 distribution of during-activation military earnings.[9]

If our hypothesis that some respondents to the 2005 SOFS-R reported annual rather than monthly totals is correct, one approach to adjusting the 2005 SOFS-R military earnings data would be to convert totals we identify as annual to monthly averages by dividing by 12. We identify annual totals as those 2005 earnings totals that lie above the 96.5th percentile of the 2004 distribution of pre-activation military earnings and those that lie above the 99.5th percentile of the 2004 distribution of during-activation military earnings.[10] We adjust 21 per-

---

[6] The question in 2005 instructed respondents to exclude reenlistment bonuses and imminent danger/hostile fire pay. If anything, this change in wording would lead earnings reported in the 2005 SOFS-R to be lower than earnings reported in the 2004 SOFS-R. Note that similar wording changes occurred in the questions on civilian earnings. The consequences for this wording change are similar to those for military earnings and are discussed below.

[7] A similar type of misreporting occurs in the Census "Long Form," where respondents give daily hours instead of usual weekly hours (Baum-Snow and Neal, forthcoming, 2008).

[8] The questions on earnings before activation appear first in the questionnaire sequence.

[9] Another interpretation that is consistent with this pattern is that respondents who have high earnings before activation will also have high earnings during activation. However, this interpretation is somewhat implausible, because the calculation of the 99th percentile of the 2004 distribution was done separately by pay grade grouping (the survey identifies five rank groupings: E1-E4, E5-E9, W2-W5, O1-O3, and O4-O6), and much of the heterogeneity in true military earnings is absorbed by pay grade.

[10] These cutoffs were chosen to minimize a measure of the difference between the 2004 and adjusted 2005 SOFS-R distributions, specifically, to minimize the maximal difference between the cumulative distribution functions of the 2004 and adjusted 2005 SOFS-R earnings distributions. Formally, define f(x) to be the cumulative distribution function for 2004

cent of 2005 SOFS-R pre-activation military earnings totals and 5 percent of 2005 SOFS-R during-activation military earnings totals in this way.

Although this approach is ad hoc, it offers a straightforward way to assess the importance of the misreporting of military earnings that appears to have occurred in the 2005 SOFS-R. The two lower panels of Figure 3.1 show the unadjusted 2004 and adjusted 2005 SOFS-R military earnings distributions. To ease visual comparisons, a log scale is used for the horizontal axis.[11] The spike in the 2005 distribution in the right tail still exists after making the adjustment, but it is much smaller than it was before this simple data edit.

Taken together, these analyses suggest that some fraction of the 2005 SOFS-R respondents reported annual rather than monthly military earnings, a consequence we believe of a small change to the wording of the earnings questions. This misreporting appears to have resulted in some very large values for monthly military earnings in the 2005 SOFS-R. In our view, this misreporting seriously undermines the usefulness of the 2005 SOFS-R for understanding how the earnings of reservists change when they are activated.

Difference in Estimates of Earnings Changes Attributable to Civilian Earnings

We employ a similar approach to quantifying the importance of differences in estimates of civilian earnings as we do for quantifying the importance of differences in estimates of military earnings. However, in the case of civilian earnings, our analyses are limited by the fact that the administrative data record annual, not monthly, earnings.

Aligning the Civilian Earnings Definitions

For the period before activation, we compare the survey estimates (average monthly civilian earnings before the most recent activation) with average monthly civilian earnings recorded in the administrative data for the calendar year preceding the year in which the respondent's most recent activation began. In this way, our administrative data estimate of pre-activation monthly civilian earnings is not biased by earnings received while activated. However, we recognize that the pre-activation period as we define it in the administrative data still might not coincide with the pre-activation period the respondent had in mind when answering the survey questionnaire.

Our analysis of differences in civilian earnings during activation is limited to the small proportion of respondents whose period of activation spanned an entire calendar year (10 percent of respondents in 2004 and 21 percent of respondents in 2005). Consider, for example, a reservist who was activated in August 2003 and deactivated in March 2004. We cannot determine what this individual's civilian earnings were during that period of active-duty service in the administrative data, since those data are recorded annually. But we can estimate civilian earnings during activation in the administrative data for a reservist who was activated for all of

---

earnings and $g(x,c)$ to be the density for 2005 adjusted earnings. Adjusted earnings, x, are defined to be observed earnings, x^*, if $x^* < c$ and equal to $x^*/12$ if $x^* > c$. The cutoff, c, was chosen to minimize the maximum (over x) of $|f(x) - g(x,c)|$. Note that this is the statistic used in the Kolmogorov-Smirnov test that two independent samples are drawn from the same distribution (Conover, 1999).

11 To use a log scale, we exclude observations with zero earnings. The percentage reporting zero earnings before activation is 5.8 percent in 2004 and 3.4 in 2005. For earnings during- activation, 2.8 percent of observations are zero in 2004 and 3.5 percent in 2005.

Figure 3.1
Comparison of 2004 and 2005 SOFS-R Military Earnings Distributions

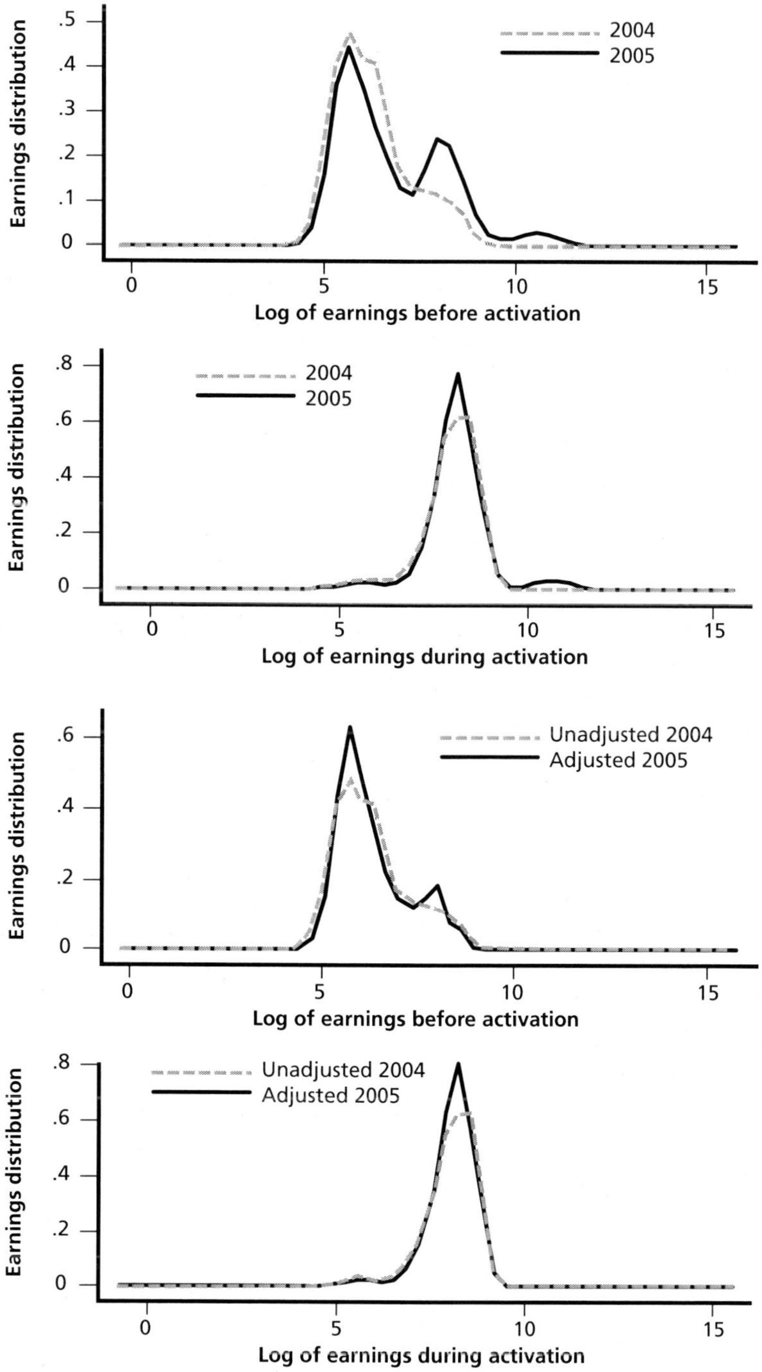

RAND *TR565-3.1*

2003 or for all of 2004. Thus, our civilian earnings comparisons generalize best to the population of reservists activated for an entire calendar year.

Table 3.5
2004 and 2005 SOFS-R Military Earnings Distributions (in dollars)

	Mean	Median	75th Percentile	90th Percentile
A. Earnings Before Activation				
2004 unadjusted	712	300	536	1,666
2005 unadjusted	4,136	450	2,500	4,665
2005 adjusted	885	333	750	2,400
B. Earnings During Activation				
2004 unadjusted	2,932	2,700	3,999	5,200
2005 unadjusted	5,917	3,000	4,200	6,400
2005 adjusted	3,183	2,950	4,000	5,500

Differences in SOFS-R and Administrative Estimates of Civilian Earnings

Table 3.6 compares the civilian earnings of reservists as reported in the SOFS-R and administrative data employing the methods just described. We focus on the results for 2004, since the same issues that affected military earnings in the 2005 SOFS-R also appear to have affected reports of civilian earnings in that year (this issue is discussed further below). Our baseline estimates show that average civilian earnings before activation are substantially higher ($823 or 36 percent higher) in the SOFS-R than in the administrative data. Conversely, during activation, civilian earnings are considerably higher ($736 or 68 percent higher) in the administrative data than in the SOFS-R.

However, these baseline estimates are likely influenced by the misalignment of civilian earnings concepts in the two data sources. In particular, because reservists who return from activation in the middle of the year will usually reenter the civilian labor market, average monthly civilian earnings received in the year during activation are likely to overstate civilian earnings received during activation. Similarly, average civilian earnings received in the year before the most recent activation would understate average earnings received in the period immediately before the most recent activation if the pre-activation year included a second earlier activation that would draw reservists out of the civilian labor market.

Our strategy for aligning the earnings definitions is to limit our analysis to the subsample of reservists who were not activated during any part of the pre-activation year and whose activations then spanned an entire calendar year. In this restricted sample, civilian earnings during activation are still higher in the administrative data, but the difference is much smaller ($264 compared with $736 in the baseline sample). In percentage terms, the difference in pre-activation earnings between the survey and administrative estimates is about the same as the baseline difference, 37 percent.

Employing this restricted sample to assess differences in civilian earnings has the disadvantage of generating estimates of differences that do not necessarily generalize to the reserve population at large. For example, 61 percent of the full 2004 SOFS-R sample (weighted) consists of senior personnel (E5-E9 or O4-O6); this compares with 81 percent (weighted)

Table 3.6
Comparison of Mean Civilian Earnings in SOFS-R and Administrative Data (in dollars)

Sample	SOFS-R	Administrative Data	SOFS-R Minus Administrative Data	% Difference	Number
A. 2004, Earnings Before Activation					
All	3,078	2,255	823	36	7,720
No prior activation	3,990	3,100	890	29	6,218
No prior activation and activation spans calendar year	4,097	2,995	1,102	37	758
B. 2004, Earnings During Activation					
All	342	1,078	–736	–68	7,720
No prior activation and activation spans calendar year	514	778	–264	–34	758
C. 2005, Earnings Before Activation					
All	8,083	2,112	5,970	283	33,296
No prior activation	8,955	2,663	6,292	236	27,390
No prior activation and activation spans calendar year	10,113	2,494	7,619	306	7,183
D. 2005, Earnings During Activation					
All	487	905	–418	–46	33,296
No prior activation and activation spans calendar year	517	584	–67	–12	7,183

NOTE: Percentages are calculated relative to administrative data totals.

in the restricted 2004 SOFS-R sample.[12] The restricted sample might also differ in terms of characteristics we do not observe. For instance, the restricted sample consists of individuals whose activations last at least one year and it may be that these individuals have particularly strong preferences for activation, perhaps because their civilian labor market opportunities are poor. Thus, the selected nature of the restricted sample makes it difficult to draw firm conclusions about the degree of misreporting in civilian earnings in the overall SOFS-R sample.

Why Do SOFS-R and Administrative Estimates of Civilian Earnings Differ?

There are at least four reasons why the 2004 SOFS-R might generate higher estimates of civilian earnings before activation than what is recorded in the administrative data. One possibility is that SOFS-R respondents overreport civilian earnings. Although this is certainly possible, most of the evidence on misreporting indicates that survey respondents typically underreport income, partly because they omit certain sources of income (Bound, Brown, and Mathiowetz, 2001; Roemer, 2000; Nelson et al., 1998).[13]

---

[12] If the results are reweighted to account for differences in pay grade in the restricted sample, the differences are somewhat larger ($460).

[13] This type of underreporting could occur in this case if respondents hold two civilian jobs but report the earnings from only one of them.

However, there is some published evidence of overreporting of civilian earnings. Bollinger (1998) found evidence that in the CPS, low earners overreport wage income, as did Rodgers, Brown, and Duncan (1993) in the Panel Study of Income Dynamics. Similarly, an evaluation of the federal Job Corps program found that earnings were likely overreported (Schochet, McConnell, and Burghardt, 2003).[14] Since many reservists, especially junior enlisted service members, have fairly low earnings, these studies may be pertinent to the reporting tendencies of reservists and suggest that overreporting of civilian income in the SOFS-R is certainly a possibility.[15]

A second possibility is that SOFS-R respondents include asset and other nonwage income whereas the SSA data include only labor market earnings. This would be an example of earnings definition misalignment rather than of misreporting, since the 2004 SOFS-R civilian earnings questions do not explicitly ask respondents to exclude income from other sources.[16] Given that most reservists are fairly young and not especially wealthy, however, it seems unlikely that the inclusion of nonwage income in the 2004 SOFS-R estimates could explain a large portion of the difference between SOFS-R and administrative data.

Third, the time period that respondents have in mind might be different from the calendar year that we use to observe earnings in the administrative data. This is almost certainly true to some extent, but it is not clear what direction or how large the bias this type of earnings concept misalignment would impart.

A final possibility is that SOFS-R respondents report labor market earnings not recorded in the administrative data. As discussed in Chapter Two, almost all labor market earnings should be reported to SSA, but sometimes they are not. And since reservists might be more likely to hold irregular jobs (that is, jobs that are temporary and for which employers do not report earnings to the IRS (illegally)) in the informal labor market, incomplete coverage of total civilian earnings on the MEF records might be especially problematic.

Assessing the importance of the underreporting of earnings to SSA is difficult. Blakemore et al. (1996) report that employers fail to report earnings for more than 13 percent of employees eligible for unemployment insurance (UI), and this figure does not account for under-the-table earnings.[17] Since reservists might be more likely to have temporary, irregular civilian employment, or to work as independent contractors, the MEF records might miss an especially large share of income for this population. On the other hand, a study of the Job Corps program found that lack of accounting for both formal and informal jobs in administrative earnings data (specifically, UI records) cannot explain much of the difference between survey and administrative reports of earnings. Furthermore, coverage is higher for SSA Medicare earnings

[14] This overreporting appears to be due to overreporting of hours worked (information on earnings was not asked directly but was computed by multiplying the hourly wage by hours worked). It is not clear whether asking directly about earnings—as is done in the SOFS-R—would have improved the accuracy of the information that was collected.

[15] This type of overreporting could occur if respondents exclude periods of unemployment from the computation of average earnings.

[16] The 2005 survey question instructs respondents to report only civilian earnings income and explicitly asks them to exclude income from other sources.

[17] As with taxes to fund Medicare, employers are also required to pay taxes on the earnings paid to employees who are covered by a state's UI system. UI earnings records submitted by employers are frequently used as a source of information on labor market earnings (Jacobson, LaLonde and Sullivan, 1993). Fewer jobs are covered by UI than are subject to Medicare taxes.

than for UI earnings (government employees and the self-employed are in the SSA Medicare earnings data but not in the UI data).

It is instructive to consider why these alternative explanations for differences between SOFS-R and administrative estimates of civilian earnings are less relevant for understanding differences in estimates of military earnings. First, the military earnings questions do not ask respondents to exclude any types of military pays or allowances, so the appropriate earnings definition is total military compensation. If respondents report only certain types of pay (such as basic pay), that should be considered misreporting.[18] Second, military pay records cover all forms of compensation. Third, the availability of monthly military pay data allows us to examine average earnings over several time intervals that respondents could have had in mind when answering the military earnings questions.

Finally, we note that none of these explanations can explain why the administrative data imply somewhat higher earnings during activation than do the SOFS-R data. One possibility is that respondents who received earnings from their civilian employers while activated (a small percentage of civilian employers continue to pay reservists part or all of their civilian wages while they serve on active duty) nonetheless reported zero civilian earnings.

The results in Table 3.4 suggest that the omission of the tax advantage and misreporting of military earnings in the 2004 SOFS-R can explain as much as 70 percent of the difference in baseline estimates of earnings changes derived from SOFS-R and administrative data sources. Although this represents a substantial proportion of the total baseline difference, a large gap in the estimates remains. Much of the gap is likely due to differences in reports of civilian earnings.[19] Ideally, we would like to quantify how much of the baseline difference is attributable to misreporting civilian earnings in the SOFS-R using the approach we employed to quantify the importance of misreporting military earnings. However, because aligning the civilian earnings definitions in the two data sources is so difficult, we do not estimate a specific percentage of the baseline difference that is accounted for by misreporting civilian earnings.

Explaining the Difference Between the 2004 and 2005 SOFS-R Estimates of Civilian Earnings

As with military earnings, the 2005 SOFS-R implies much higher civilian earnings than does the 2004 SOFS-R. It is likely that changes in question wording also can explain this result. The 2005 SOFS-R question about civilian earnings before activation instructed respondents to provide an average over the 12 months before the most recent activation. The 2004 SOFS-R question asked for a monthly average over an unspecified period of time. We believe that this wording change prompted some respondents to report 12-month totals instead of 12-month averages, leading to a number of unusually large values for monthly civilian earnings in the 2005 SOFS-R. This can be seen in Figure 3.2, which plots the distribution of earnings before

[18] The 2005 survey items instruct respondents to exclude bonus payments; we do not include bonus payments in any calculations in this study.

[19] Some of the remaining discrepancy might be due to total earnings concept misalignment. Recall that the baseline administrative data estimates refer to average earnings received in the activation year, which potentially includes earnings both before and after activation. The survey, in contrast, asks about earnings in the activation year. Thus, even if the SSA earnings data captured all civilian earnings and there were no misreporting, the survey and administrative data earnings change estimates might not coincide perfectly.

Figure 3.2
Comparison of 2004 and 2005 SOFS-R Civilian Earnings Distributions

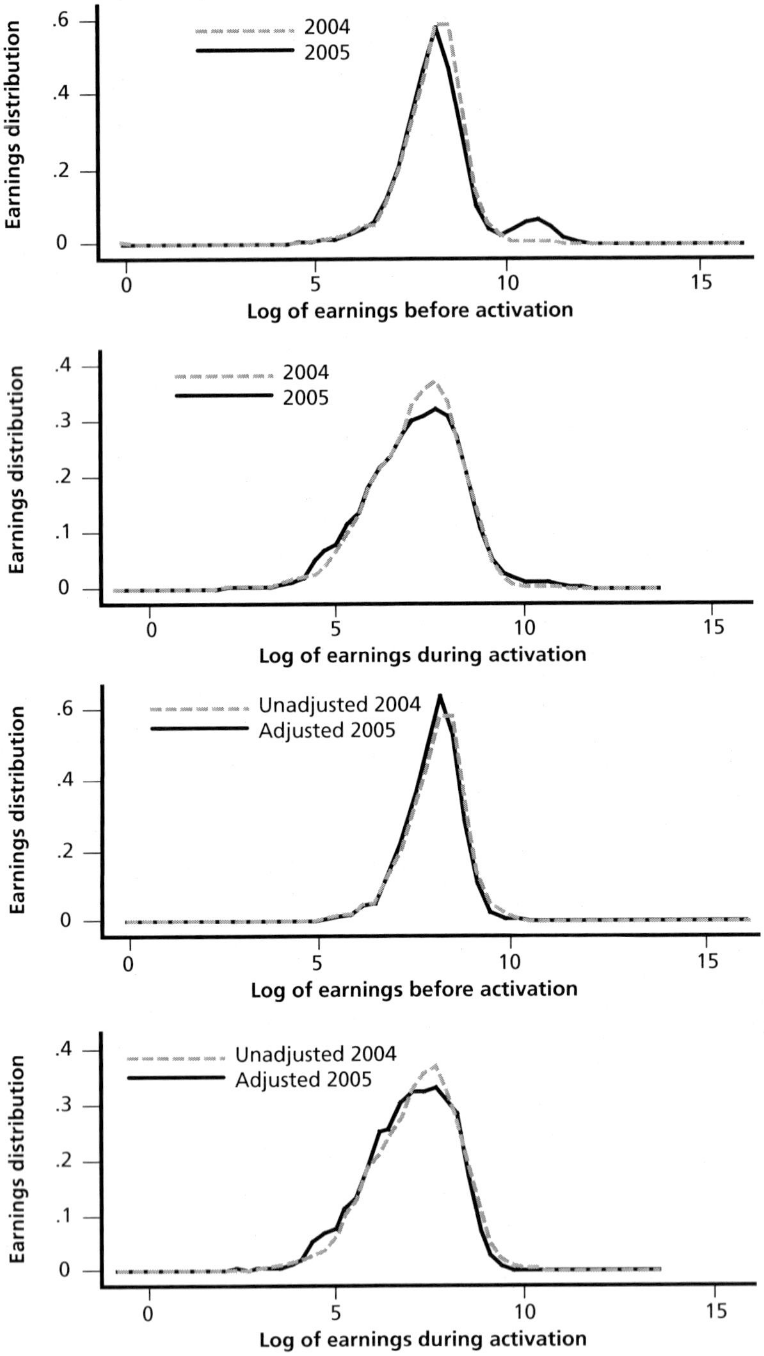

RAND *TR565-3.2*

and during activation recorded in the 2004 and 2005 SOFS-R. Again a log scale is used to help make visual comparisons.[20] The distributions of earnings before activation are similar except in the upper tail. The question wording about civilian earnings during activation remained basically unchanged and we see in Figure 3.2 that the 2004 and 2005 distributions of civilian earnings during activation are quite similar.

We edited the 2005 SOFS-R civilian earnings data by converting values that appeared to be annual totals to monthly averages. We identified these values employing the 2004 SOFS-R civilian earnings distribution and the method described above for adjusting 2005 SOFS-R military earnings totals. After making this adjustment to the 2005 SOFS-R civilian earnings data, the 2004 and 2005 SOFS-R distributions of civilian earnings before activation are virtually identical. As seen in Table 3.7, the mean of the adjusted 2005 data is $293, which is slightly less than the mean in 2004 ($379).

Table 3.7
2004 and 2005 SOFS-R Civilian Earnings Distributions

	Mean	Median	75th Percentile	90th Percentile
A. Earnings Before Activation				
2004 unadjusted	2,956	2,500	4,000	5,500
2005 unadjusted	7,633	2,500	4,406	8,500
2005 adjusted	3,031	2,500	4,000	5,700
B. Earnings During Activation				
2004 unadjusted	379	0	0	1,350
2005 unadjusted	484	0	0	1,000
2005 adjusted	293	0	0	950

[20] In 2004, 9.1 of observations have zero earnings compared with 10.6 percent in 2005. For earnings during activation, the fraction reporting zero is equal to 80.6 in 2004 and 82.9 percent in 2005.

CHAPTER FOUR

Analysis of Nonresponse Bias

The previous chapter assessed the importance of differences in earnings reports for the subsample of individuals who responded to the SOFS-R and who were matched to administrative data. However, the sample employed by LKM is not conditional on responding to the SOFS-R. This opens up the possibility that some of the difference between the estimates of earnings changes reported in LKM and those derived from the SOFS-R are due to differences in the characteristics of individuals who responded to the SOFS-R.

This issue would not be a major concern if the SOFS-R response rates were not so low. Effective response rates in the 2004 and 2005 SOFS-R are 34 and 30 percent, respectively. Moreover, as we show below, these response rates differ sharply by pay grade, which suggests that survey nonresponse could help explain differences between the LKM and SOFS-R estimates of earnings changes. This chapter assesses the substantive effect of nonresponse bias in the SOFS-R.

An Approach to Quantifying Nonresponse Bias in the SOFS-R

To understand our analysis strategy, it is crucial to understand the different ways in which survey nonresponse could bias estimates of earnings changes. Suppose that earnings changes differ across the population of reservists, as shown in LKM. If the determinants of survey nonresponse are correlated with earnings changes, then the earnings patterns of survey respondents will differ systematically from those of survey nonrespondents. We refer to the difference between the estimate using only respondents and the estimate that would be obtained using both respondents and nonrespondents as nonresponse bias.

When faced with survey nonresponse, the standard correction is to reweight the data for respondents so that the reweighted data resemble the entire sample—both respondents and nonrespondents (Groves, 2006; Bethlehem, 2002). Heuristically, the adjusted weights give more (less) importance to responses from individuals in subgroups that are relatively less (more) likely to respond to the survey. DMDC's survey weights make this adjustment.[1] However, if

[1] The sampling design stratified the target population on the basis of (1) reserve component (Army National Guard, Army Reserve, Naval Reserve, Marine Corps Reserve, Air National Guard, and Air Force Reserve), (2) gender, (3) pay grade (E1–E3, E4, E5–E6, E7–E9, W1–W5, O1–O3, and O4–O6), (4) reserve program (TPUs (Troop Program Units)/unknown, Active Guard and Reserve/Training and Administration of the Reserve, Military Technicians, and Individual Mobilization Augmentee), (5) ethnicity (nonminority, minority), and (6) whether the reservist had been activated in the prior 24 months. The survey weights were computed based on these stratifying variables.

respondents and nonrespondents differ within observable subgroups, reweighting based on observable characteristics will not correct the bias that results from differential nonresponse.

The presence of such differential nonresponse (even after reweighting) and the bias it imparts to survey data are major areas of research in the survey literature. The recent literature on this topic suggest that nonresponse bias is generally less important than had once been feared (Keeter et al., 2000; Curtin, Presser, and Singer, 2000; Merkle and Edelman, 2002). However, other research has found evidence of substantively important nonresponse bias (Schochet, McConnell, and Burghardt, 2003; Olson, 2006).[2]

The existing literature addresses civilian surveys that are typically based on population samples. These surveys can reweight based only on gross population characteristics such as gender, age, and race/ethnicity. In contrast, military surveys are based on list samples. The list comes from the military personnel files. Those files include detailed information about the sample including not only gender, age, and race/ethnicity but also pay grade and component. Pay grade is a particularly useful covariate that is likely to be a strong predictor of many outcomes, including earnings. Therefore, we would expect reweighting to be more effective in reducing nonresponse bias in military surveys than in civilian surveys.

Consistent with this discussion, we began by comparing unweighted average military and civilian earnings of respondents and nonrespondents, as recorded in the administrative data, to test whether nonresponse in the SOFS-R is correlated with earnings levels and changes. We then make those same comparisons but employing the SOFS-R weights for SOFS-R respondents. If the weights eliminate all nonresponse bias, then the weighted survey estimates should be very close to estimates for the entire sampling frame. The difference between the weighted survey estimates and the estimates from the sampling frame is our estimate of nonresponse bias. This approach has been used by other researchers trying to gauge the severity of nonresponse bias (Olson, 2006; Bolstein, 1991).[3]

Estimates of Nonresponse Bias in the SOFS-R

We begin by comparing the pay grade distribution and mean administrative data earnings of SOFS-R respondents and nonrespondents, where nonrespondents are reservists who were sampled but either did not complete a survey or did not provide valid responses to the earnings questions. The results in Table 4.1 show clear evidence of systematic nonresponse in both the 2004 and 2005 SOFS-R.[4] Almost half of the SOFS-R nonrespondents are junior enlisted reservists (E1–E4) compared with only 19 percent of SOFS-R respondents. Conversely, SOFS-R

---

[2] See Groves (2006) for a thorough summary of the literature on nonresponse bias. His meta-analysis suggests that nonresponse bias exists but that response rates are not predictive of the magnitude of nonresponse bias.

[3] Our analysis examines the nonresponse bias of mean earnings levels and changes. Nonresponse might generate biases of other statistics that we do not consider here. Similarly, these analyses will be informative only about whether nonresponse leads to biased estimates that use administrative data. It may be that nonresponse bias is different for estimates that use earnings from the survey itself or for any other survey items (Groves, 2006).

[4] These deviations are so large as to be clearly statistically significant. They are clearly so using simple chi-square tests. Those tests are not formally appropriate given the complex nature of the SOFS-R sampling design. However, the rejections are so resounding that it seems unlikely that even the corrected tests would fail to reject.

Table 4.1
Pay Grade and Administrative Earnings, by SOFS-R Respondent Status (Unweighted)

	2004 SOFS-R			2005 SOFS-R		
	Nonresponse	Unweighted Response	% Difference	Nonresponse	Unweighted Response	% Difference
A. Pay Grade Distribution (in percent)						
E1–E4	48.6	18.7	160	45.2	17.5	158
E5–E9	34.9	38.2	−9	45.5	53.6	−15
W1–W5	1.8	4.6	−60	0.9	2.7	−67
O1–O3	5.8	12.6	−54	3.9	9.7	−59
O4–O6	8.8	26.0	−66	4.6	16.5	−72
B. Mean Annual Earnings (in dollars)						
Before activation						
Civilian	22,430	36,828	−39	19,806	31,499	−37
Military	10,248	14,802	−31	8,750	13,134	−33
Total	32,678	51,630	−37	28,556	44,633	−36
During activation						
Civilian	12,170	20,338	−40	8,904	14,002	−36
Military	31,459	43,378	−27	32,631	44,493	−27
Total	43,629	63,711	−32	41,535	58,496	−29
Earnings change	10,951	12,080	−9	12,979	13,863	−6
Number	7,720	21,964		123,700	33,296	

NOTE: The percentage difference is given relative to the (unweighted) respondent average.

respondents are considerably more likely to be officers than are SOFS-R nonrespondents (39 percent compared with 15 percent).[5]

Since nonrespondents are less likely to be officers and more likely to be fairly junior, it is not surprising that their earnings are substantially lower than those of respondents. Average total earnings in the year before activation are 37 percent higher among respondents than among nonrespondents in 2004, and 36 percent higher in 2005. Earnings in the year of activation are about 30 percent higher among respondents.

For most purposes, differences in earnings of this magnitude between respondents and nonrespondents are likely to matter a great deal. However, differences in earnings levels might not translate into differences in earnings changes. The average earnings change among nonrespondents is within 6 percent of the earnings change among respondents for the 2004 SOFS-R

[5] These comparisons are unweighted. In particular, they do not adjust for the intentional proportional sampling design whereby certain groups (such as senior officers) were oversampled. However, the comparisons represented in Table 4.1 are made conditional on having been sampled. With random nonresponse, the pay grade distributions should be the same among respondents and nonrespondents even though nonproportional sampling implies that the distribution of pay grade in the population will not be the same as the unweighted distribution in the sample.

and within 9 percent of the earnings change among respondents for the 2005 SOFS-R. This is because differences in earnings levels before and during activation are in the same direction and of a similar magnitude and so the influence of nonresponse is effectively differenced out in the case of earnings changes.[6]

Of course, earnings levels are frequently the object of interest. To assess the substantive importance of nonresponse bias in earnings levels, we compare the weighted average earnings of respondents with the average earnings of the overall sampling frame. The weighted distribution of pay grade among survey respondents reported in Table 4.2 is similar to that of the sampling frame, with officers being slightly overrepresented in the weighted survey data.

There are several reasons why the two pay grade distributions might differ. First, although the weights are designed to adjust for nonresponse to the entire survey, no adjustment is made for nonresponse to specific survey items. In contrast, we treat individuals with missing or invalid earnings data as nonrespondents. Second, we did not successfully link all SOFS-R

Table 4.2
Pay Grade and Administrative Earnings, by SOFS-R Respondent Status (Weighted)

	2004 SOFS-R			2005 SOFS-R		
	Weighted Response	Sampling Frame	% Difference	Weighted Response	Sampling Frame	% Difference
A. Pay Grade Distribution (in percent)						
E1–E4	32.7	34.4	–5	32.5	32.7	–1
E5–E9	53.0	52.3	1	53.1	54.0	–2
W1–W5	1.0	1.1	–3	1.2	1.1	6
O1–O3	5.0	4.5	11	5.5	4.7	16
O4–O6	8.3	7.7	8	7.7	7.5	3
B. Mean Annual Earnings (in dollars)						
Before activation						
Civilian	27,063	25,343	7	23,743	25,347	–6
Military	10,457	10,091	4	10,338	10,090	2
Total	37,520	35,434	6	34,081	35,437	–4
During activation						
Civilian	12,941	11,752	10	10,855	10,430	4
Military	35,471	35,324	0	37,494	36,791	2
Total	48,411	47,076	3	48,348	47,221	2
Earnings Change	10,891	11,642	–6	12,979	13,863	–6
Number	7,720			33,296		

NOTE: The percentage difference is given relative to the sampling frame average.

[6] Note further that the magnitude of the difference in annual earnings change is a little over $1,000, whereas the baseline difference between the survey and administrative monthly earnings data estimate was over $1,600.

respondent records to administrative data, so even after weighting, our matched sample might not be representative of the sampling frame. Third, there is likely some discrepancy between the sampling frame used by DMDC to conduct its survey and our attempt to replicate the sampling frame using the WEX.[7] Finally, it is important to recognize that the difference between the weighted survey and the sampling frame is small. For both the 2004 and 2005 SOFS-R, the weighted proportion of reservists in a given pay grade cell never differs from the proportion in the sampling frame by more than 2 percentage points.

The earnings results in the lower panel of Table 4.2 show that mean earnings are higher among respondents, even after using the survey weights. However, the magnitude of the difference between the survey respondents and the sampling frame is modest. This is especially true for military earnings, where the weighted average in the survey is always within 4 percent of the average in the administrative data. For civilian earnings in the year before activation, average earnings differ by about 6 to 7 percent between the survey and the sampling frame. In the activation year, the difference is 4 percent in the 2005 SOFS-R and 10 percent in the 2004 SOFS-R.

The fact that the nonresponse bias is more pronounced for civilian earnings than for military earnings can be understood by noting that most military income is basic pay, which is mechanically related to a reservist's years of service and pay grade.[8] Since the survey weights closely align the distribution of pay grade in the survey with that in the sampling frame, it is not surprising that nonresponse bias for military earnings is small. In contrast, the process that determines civilian earnings is more complicated and is driven by such factors as age, education, experience, and local labor market conditions. The covariates used to construct the survey weights are likely correlated with these factors, but the strength of this correlation is likely not as strong as it is with determinants of military earnings. To the extent that unobservable differences between respondents and nonrespondents remain after adjusting for the survey weights, estimates of civilian earnings will be biased.

In summary, we find clear evidence of systematic nonresponse that generates differences across respondents and nonrespondents in average earnings recorded in administrative data. However, most of this bias is effectively differenced out when computing earnings changes. Moreover, SOFS-R survey weights eliminate most of the nonresponse bias in earnings levels. Thus, the practical significance of nonresponse for analyzing earnings changes during activation is likely to be small. This is due in part to the fact that the survey weights are based on a rich set of characteristics that are strongly related to earnings (e.g., pay grade). Nonresponse bias could be more significant for SOFS-R items that are not as strongly related to the covariates used to construct the survey weights.

---

[7] To get a sense of how much slippage there might be in our construction of the survey sampling frame, note that the information available on the WEX indicates that about 7 percent of respondents in the matched sample are not in the sampling frame. Note that we included anyone whose survey record indicated that they were eligible for the survey regardless of whether they were identified in the WEX as being in the sampling frame.

[8] Earnings heterogeneity among reservists is mainly due to differences in the amount of time spent on active duty as well as on eligibility for special pays and allowances.

CHAPTER FIVE

Conclusion

This report has given the results of a set of analyses designed to account for differences in estimates of earnings changes derived from SOFS-R and administrative data. We argue, *a priori*, that there are at least five reasons why the SOFS-R and administrative estimates might differ: (1) the administrative estimates incorporate the federal tax advantage whereas the SOFS-R estimates do not, (2) survey respondents might misreport military and civilian earnings, (3) recorded earnings in the administrative data may not include all labor market earnings, (4) the survey and administrative data earnings definitions are not identical, and (5) the SOFS-R respondent sample might be unrepresentative of the reserve population at large.

Our analyses help us to understand the relative importance of each of these reasons in explaining differences between SOFS-R and administrative estimates of earnings changes. Our preferred estimates imply that differential treatment of the tax advantage and misreporting of military earnings in the SOFS-R account for about 70 percent of the overall difference in estimates of earnings changes derived from the 2004 SOFS-R and administrative data. Although we find that civilian earnings before activation are substantially higher in the SOFS-R than in the administrative data, we cannot conclude that this difference is due to misreporting alone because of difficulties in aligning the civilian earnings definitions and the possibility that the SSA MEF earnings records omit some sources of civilian earnings.

We also find that the very large mean earnings losses in the 2005 SOFS-R are due to a small number of respondents who reported very high pre-activation earnings. These respondents appear to have misunderstood the question wording and provided annual instead of monthly earnings figures. We show that a simple adjustment to the 2005 earnings data (dividing values that appear to be annual figures by 12) produces a distribution of earnings that closely resembles the distribution in the 2004 SOFS-R.

Finally, our analyses indicate that survey nonresponse can explain little of the difference between the SOFS-R and administrative data estimates. Although SOFS-R respondents are clearly not representative of all sampled reservists in terms of pay grade and earnings, estimates of earnings changes are fairly similar and in the same direction for respondents and nonrespondents. Moreover, the SOFS-R weights go a long way toward correcting any nonresponse bias for earnings levels.

Our findings have a number of implications. First, analysts and policymakers should employ SOFS-R data on military earnings with caution. One reason is the SOFS-R earnings data do not include the value of the federal tax advantage. This issue becomes especially important when analyzing earnings during activation, since many of the pays and allowances reservists receive while activated are tax exempt. A second reason is that SOFS-R respondents appear to underreport military earnings substantially. Both the omission of the tax advantage and underreporting of military earnings lead the SOFS-R to estimate large average earnings losses rather than the large earnings gains apparent in the administrative data.

Our analyses do not permit us to draw strong conclusions about the quality of the SOFS-R civilian earnings data. The SSA MEF does not include under-the-table earnings. If such unreported income is an important component of civilian earnings before activation, then our earnings loss estimates using administrative data would be understated. On the other hand, the problems with the SOFS-R military earnings data suggest that analysts should to be cautious about using the civilian earnings data as well.

It follows that military personnel analysts should employ administrative data when feasible. Processing pre-existing administrative data is less expensive and less time-consuming than collecting comparable survey data. Furthermore, administrative data on earnings are likely to be more accurate than self-reported earnings recorded in surveys, although analysts should also be aware that administrative data can miss some sources of earnings (for example, under-the-table earnings). A significant limitation of administrative data is the relatively small amount of information they contain about the study population, including certain critical objective characteristics. And subjective data, such as reenlistment intentions, can be collected by only survey. Thus, the best option available to the analyst may often be to match administrative data on key objective characteristics to survey data containing a richer array of respondent characteristics, intentions, and attitudes.

Finally, our results have methodological implications for survey data collection. We find that nonresponse bias is modest even though response rates to the SOFS-R are low. To the extent that this finding generalizes to other survey populations, this suggests that researchers designing surveys should devote more effort to minimizing misreporting rather than to improving response rates. For instance, surveys that collect earnings data should ask questions on income from a greater number of sources to prevent underreporting. On the other hand, these findings apply to a survey where factors that are strongly related to earnings (in particular, pay grade) are observed in the sample population and can be used to adjust for differential nonresponse. In other surveys where survey weights are based on factors that are not as strongly related to the survey variable of interest, nonresponse bias could be more problematic.

APPENDIX A

Administrative Data Estimates of Changes in Reserve Earnings Attributable to Activation

This appendix reports estimates of the effect of activation on the earnings of reservists following the approach of LKM. These estimates extend LKM in three ways. First, LKM report results for activations occurring in 2002 and 2003. Here, we report estimates for activations occurring in 2004 and 2005 as well. Second, because of data problems at DMDC and SSA, LKM were unable to find earnings records for about 15 percent of the reservists sampled in that study. Those data problems have now been largely resolved and the number of reservists with missing earnings records is far fewer. Finally, it is possible that some specific military occupations are more likely than others to experience earnings losses as a result of activation. To address this possibility, we present estimates of earnings changes attributable to activation by one- and three-digit military occupation.

Overall, the results of these new analyses reveal the same basic patterns as reported in LKM: large average earnings gains and relatively rare earnings losses. These results hold across activation years, activation duration, pay grade, and military occupation. The remainder of this appendix has the following structure. The first section describes how we constructed our administrative dataset. The second section describes our methods for estimating earnings changes attributable to activation. The third section then presents estimates of earnings changes by year, active-duty days served, pay grade, and military occupation.

Data and Sample

To estimate how activation affects earnings, we employ data on civilian earnings, military earnings, and time on active duty. In the most general terms, we construct this database by merging information on time on active duty and military allowances from DMDC administrative data and information on civilian earnings and military pay from SSA. Additionally, we use this information to impute a value to the federal tax preference accorded to some military compensation.

Data Sources

We define earnings in this report as real annual after-tax equivalent cash compensation. We approximate this concept as the sum of four components:

- *Civilian earnings*: Civilian earnings include all nonmilitary earnings subject to Medicare taxes. We obtain data on annual earnings from the SSA MEF. SSA uses earnings data recorded in the MEF to compute Social Security benefits and to compute Social Security and Medicare taxes.

- *Military pay*: Military pay includes all military pays (e.g., basic pay, hostile fire pay) and bonuses. These pays are reported directly to SSA to compute Social Security benefits and Medicare taxes and are included in the earnings measure we obtain from the MEF.
- *Military allowances*: Military allowances include all military allowances (e.g., Basic Allowance for Subsistence, Basic Allowance for Housing, Family Separation Allowance). Allowances are computed from DMDC's Reserve Pay File (for all unactivated reservists and activated Army and Air Force reservists) and from DMDC's Active Duty Pay File (for activated Navy and Marine Corps reservists).
- *Tax advantage*: Military allowances and all military pays received while serving in a combat zone are not subject to federal income taxes. To allow for a consistent comparison of earnings when activated and not activated, we use tax tables to impute taxable-equivalent earnings. The tax imputations assume that the reservist files as single with no dependents and account for all federal income taxes and Social Security taxes. The imputations do not account for state taxes.

The SSA data are available on a calendar-year basis, so our analyses consider only annual data. All dollars are converted to $2004 using the Consumer Price Index.

Active-duty days are computed by dividing active-duty pay received by the daily basic pay amount from the current pay table given observed rank and years of service. Active-duty days include active-duty training (usually two weeks during the summer), but not inactive-duty training (IADT) (usually two days per month).

Sample

Our initial sample consists of 1,572,334 reservists who, according to the DMDC's WEX), were members of a reserve component other than the Coast Guard at any time during the period January 1999 to November 2003. We drop 98,797 reservists from this sample who do not appear in the ADPF or RPF during our sample period and 8,647 reservists for whom there was no corresponding SSA earnings record during our sample period.[1]

Data Processing and Sample Sizes

Our basic unit of analysis is an annual observation on each reservist in our sample between 2001 and 2005. Using programs we provided to SSA, SSA staff computed the difference between earnings in an out year and a base year for each reservist. The years 2001–2004 serve as out years and the years 2002–2005 serve as base years. To preserve confidentiality, the programs generate group-level statistics (e.g., mean, median) on these earnings differences.[2]

The groups are defined by all combinations of the following four variables:

- *active-duty days in the base year*: 0–30, 31–90, 91–180, 181–270, and 271 active-duty days or more

[1] In addition, we drop reservists who served in the regular active-duty force in the year before serving on active duty or had a rank of more than O6.

[2] Following SSA policy, statistics are generated only for groups containing five or more observations. SSA will not release data on individuals and so all analyses of individual data occurred at SSA.

- *active-duty days in the out year,* grouped as above
- *component:* Army Reserve, Army National Guard, Air Force Reserve, Air National Guard, Naval Reserve, and the Marine Corps Reserve
- *paygrade:* junior enlisted (E1–E4), senior enlisted (E5–E9), warrant officers (W1–W5), junior officers (O1–O3) and senior officers (O4–O6).

Table A.1 reports sample sizes for reservists serving 0–30 active-duty days in a given base year and different lengths of time in a given out year. As can be seen, sample sizes are reasonably large in all cases.

Table A.1
Sample Sizes, by Base Year and Out Year and Active-Duty Days Served in the Out Year

	Active-Duty Days in Out Year				
Base Year/Out Year	0–30	31–90	91–180	181–270	271+
2001/2002	460,866	51,729	28,351	13,283	9,144
2002/2003	348,446	55,211	38,939	28,535	62,075
2003/2004	290,221	38,456	31,676	20,281	19,621
2004/2005	270,807	41,526	19,121	9,270	12,408

NOTE: Samples are restricted to reservists serving 0–30 active-duty days in the base year.

We also report results by one- and three-digit military occupational codes as defined by DMDC's Occupational Database. Each servicemembers' military occupation is reported in the WEX.

Methods

We estimate the effect of activation on the earnings of reservists using a version of the difference-in-differences framework appropriate to our very large sample sizes and in consideration of the fact that we have only indirect access to the underlying data. To motivate the difference-in-differences framework, Figure 3.1 graphs with a dashed line the notional earnings of reservists who served 0–30 active-duty days in both 2004 and 2005.[3] We refer to reservists serving less than 31 days on active duty as being unactivated. We also graph in Figure 3.1, the notional earnings of reservists who served 0–30 active-duty days in 2004 but then served more than 30 days on active duty in 2005. We refer to this group of reservists as being unactivated in 2004 and activated in 2005.

Consistent with SOFS-R evidence, Figure 3.1 plots a decrease in the earnings of activated reservists between 2004 (before activation) and 2005 (while activated). In the tables below, we refer to this difference as the gross difference in earnings. Consistent with normal earnings

[3] We adopt this 30-day cutoff to exclude regular reserve duty. Standard reserve service requires training one weekend per month (which is not considered active-duty service) and two weeks during the summer (which is considered active-duty service). Many reservists serve more than 14 days on active duty in years in which they are not truly activated. This service might include additional training or special duties (e.g., disaster relief).

growth, Figure 3.1 plots an increase in the earnings of unactivated reservists. The difference-in-differences estimate is then the difference between the change in earnings of activated reservists and the change in earnings of unactivated reservists (Δ earnings of activated reservists – Δ earnings of unactivated reservists). In the tables below, we refer to the difference-in-differences as the net difference in earnings. This difference-in-differences estimate assumes that activated reservists would have experienced the same change in earnings as did the unactivated reservists had the activated reservists not been activated. Thus, the earnings experience of unactivated reservists provides an estimate of counterfactual earnings.

The discussion thus far implicitly assumes a homogenous population, which is consistent with the conventional regression implementation of difference-in-differences. That conventional difference-in-differences approach typically assumes a homogeneous effect of activation across groups defined by length of activation in the out year, paygrade, and component.

Given our large sample sizes, we do not need to maintain the assumption of homogeneous treatment effects. Instead, we compute difference-in-differences estimates for each group defined by paygrade and component.[4] We then report total difference-in-differences estimates as the weighted average of the estimates for each group, where the weights are given by the distribution of activated individuals across the groups.

The difference-in-differences approach produces an unbiased estimate of the effect of activation on earnings so long as any differences in the characteristics of reservists who are and are not activated—beyond paygrade and component—that affect earnings are fixed over time. For example, suppose that reservists who have poor civilian labor market opportunities are more likely to volunteer for active-duty service than reservists who have good civilian labor market opportunities. So long as this unobserved difference in civilian labor market opportunities is fixed in time, the difference-in-differences estimate will produce an unbiased estimate of the effect of activation on earnings. However, suppose instead that reservists who anticipate suffering declines in civilian earnings next year are more likely to volunteer for active-duty service this year. In that case, the difference-in-differences estimate could underestimate (overestimate) the positive (negative) effect of activation on earnings.

Results

We present the results of these analyses in Tables A.2 through A.7. Table A.2 presents estimates of gross and net differences in earnings by base and out year. These estimates are consistent with results reported in LKM. Table A.3 presents estimates of the fraction of reservists whose earnings fall between base and out years, the fraction whose earnings fall by more than $10,000, and the fraction whose earnings fall by more than 10 percent (gross losses). The table also reports difference-in-differences estimates of these earnings losses (net losses). Once again, the results here are consistent with those reported in LKM. Tables A.4 and A.5 report estimates of gross and net earnings changes and losses by active-duty days served in the out year and by rank for reservists activated 0–30 days in 2004 and 31 or more days in 2005. These results, too, are broadly consistent with those reported in LKM.

---

[4] Thus, for example, the earnings experiences of unactivated junior enlisted members of the Army National Guard provide the estimate of counterfactual earnings for activated junior enlisted members of the Army National Guard.

Table A.2
Gross and Net Earnings Differences, by Base and Out Year

	Earnings		Gross Difference		Net Difference	
Base Year/Out Year	**Base Year**	**Out Year**	**Level**	**% Change**	**Level**	**% Change**
2001/2002	34,546	44,685	10,138	33	8,981	29
2002/2003	35,895	47,605	11,710	38	10,960	34
2003/2004	35,284	48,959	13,675	45	12,333	40
2004/2005	36,462	50,445	13,983	46	13,208	42

NOTES: Samples are restricted to reservists serving 0–30 active-duty days in the base year. Figures are weighted according to the paygrade, component, and active-duty-days distribution of reservists serving 0–30 days in 2002 and 31 or more days in 2003. All figures are reported in $2004.

Table A.3
Gross and Net Earnings Losses, by Base and Out Year (in percent)

	Gross Loss		Net Loss	
Base Year/Out Year	**Any**	**> 10 Percent**	**Any**	**> 10 Percent**
2001/2002	16	10	–23	–14
2002/2003	15	8	–30	–18
2003/2004	12	7	–29	–19
2004/2005	14	7	–34	–21

NOTES: Samples are restricted to reservists serving 0–30 active-duty days in the base year. Figures are weighted according to the paygrade, component, and active-duty-days distribution of reservists serving 0–30 days in 2002 and 31 or more days in 2003. All figures are reported in $2004.

Table A.4
Gross and Net Earnings Differences and Losses, by Number of Active-Duty Days in 2005

	Gross Difference		Net Difference		Any Loss	
Number of Active-Duty Days	**Level**	**% Change**	**Level**	**% Change**	**Gross**	**Net**
31–90	3,841	11	3,579	9	25%	–25
91–180	9,621	26	9,359	24	15%	–35
181–270	16,051	43	15,789	41	10%	–40
271+	24,344	68	24,082	66	7%	–43

NOTES: Samples are restricted to reservists serving 0–30 active duty days in 2004. Figures are weighted according to the paygrade and component distribution of reservists serving 0–30 days in 2004 and 31 or more days in 2003. All figures are reported in $2004.

Table A.6 reports estimates of gross and net earnings changes and losses by rank group (enlisted and officer) and by one-digit military occupation for reservists serving 0–30 days in 2004 and 31 days or more in 2005.[5] The table shows that the percentage of enlisted members with any earnings loss is greatest for those classified as "Non-Oocupational," "Health care specialists," "Other technical or allied specialists," or "Electronic equipment repairers." Among

[5] These results are qualitatively similar for different base and out year combinations.

Table A.5
Earnings Differences and Losses, by Rank in 2005

	Gross Difference		Net Difference		Any Loss	
Rank	Level	% Change	Level	% Change	Gross	Net
E1–E4	9,591	45	7,726	36	18%	–26
E5–E9	10,456	26	10,975	27	16%	–37
W1–W5	15,416	24	19,594	30	13%	–45
O1–O3	16,761	28	14,346	24	15%	–30
O4–O6	20,503	24	21,737	25	15%	–40

NOTES: Samples are restricted to reservists serving 0–30 active duty days in 2004. Figures are weighted according to the component and active-duty-days distribution of reservists serving 0–30 days in 2004 and 31 or more days in 2003. All figures are reported in $2004.

officers, the percentage with any earnings losses is greatest among those classified as "General officers and executives, N.E.C. [not elsewhere classified]" and "Health care officers." In all cases, however, earnings changes are positive and difference-in-differences estimates of the effect of activation on earnings loss are negative, implying that activation lowers the probability of earnings loss.

Table A.7 reports estimates of gross and net earnings changes and losses by rank group (enlisted and officer) and by three-digit military occupation for reservists serving 0–30 days in 2004 and 31 days or more in 2005.[6] The table is further restricted to occupational groups in which more than 20 percent of its members experiences an earnings loss. For enlisted members, the four occupations with the largest percentage experiencing an earnings loss are "Sonar operator, general" (55 percent), "Investigations" (54 percent), "Military training instructor" (54 percent), and "Illustrating" (45 percent). Difference-in-differences analyses also imply net earnings losses for these four occupational groups. All other enlisted occupational groups have net earnings gains. For officers, the four occupations with the largest percentage experiencing an earnings loss are "Physicians" (38 percent), "Other fixed-wing pilots" (38 percent), "Research and development coordinator" (38 percent), and "Communications intelligence" (38 percent). However, difference-in-differences analyses imply net earnings gains for these and all other officer occupational groups. Note that the cell sizes in these eight occupational groups with the largest earnings losses are quite small, which limits the practical importance of these earnings losses and draws into question their statistical validity. Overall, the occupational groups listed in Table A.7 represent a relatively small fraction of all activated reservists: 18 percent of activated enlisted members and 31 percent of activated officers.

[6] These results are qualitatively similar for different base and out year combinations.

Table A.6
Gross and Net Earnings Differences and Losses, by One-Digit Military Occupation

One-Digit Occupation	Number of Observations	Earnings Difference		Any Loss, %	
		Gross	Net	Gross	Net
A. Enlisted					
Non-occupational	1,328	7,014	4,626	22	–19
Health care specialists	4,619	8,878	7,808	21	–27
Other technical and allied specialists	2,578	7,580	6,491	20	–26
Electronic equipment repairers	3,868	7,668	7,257	19	–31
Craftsworkers	4,845	8,959	8,567	18	–31
Service and supply handlers	8,293	8,940	8,104	18	–29
Functional support and administration	12,618	8,351	8,104	18	–31
Communications and intelligence specialists	3,882	10,314	8,690	17	–28
Electrical/mechanical equipment repairers	14,287	7,958	8,489	17	–35
Infantry, gun crews, and seamanship specialists	13,181	10,717	9,637	17	–30
B. Officers					
General officers and executives, N.E.C.	77	17,616	33,542	26	–35
Health care officers	2,449	10,400	10,188	26	–27
Intelligence officers	335	14,837	13,567	18	–29
Engineering and maintenance officers	1,092	14,368	13,484	16	–34
Tactical operations officers	2,742	18,499	19,005	16	–36
Scientists and professionals	880	17,710	13,281	15	–35
Non-occupational	289	13,419	9,803	13	–26
Administrators	854	16,718	16,566	12	–40
Supply, procurement, and allied officers	1,103	18,452	19,466	12	–39

NOTES: Samples are restricted to reservists serving 0–30 active-duty days in 2004 and 31 or more active-duty days in 2005. All figures are reported in $2004.

Table A.7
Gross and Net Earnings Differences and Losses, by Three-Digit Military Occupation: Occupations with Earnings Losses Exceeding 20 Percent

Three-Digit Occupation	Number of Observations	Earnings Difference		Any Loss, %	
		Gross	Net	Gross	Net
	A. Enlisted				
Sonar operator, general	20	–1,717	–4,343	55	10
Investigations	21	–4,735	–7,687	54	11
Military training instructor	13	–385	–50	54	–4
Illustrating	22	796	–149	45	1
Postal	27	6,387	6,542	37	–13
Seamanship	47	7,843	6,651	34	–13
Small boat operators	35	6,346	4,402	33	–8
Safety	35	2,837	5,681	31	–30
Biomedical science and allied health	76	5,199	3,434	31	–9
Sales store	49	8,387	8,467	31	–22
Navigators	31	7,287	7,960	29	–15
Radiology	87	6,976	5,529	29	–19
Auxiliary labor, general	8	6,509	5,731	29	–17
Analysis	37	7,526	4,257	29	–17
Central office	112	6,124	5,236	27	–23
Training devices	13	2,533	2,735	27	–33
Air traffic control radar	52	1,385	1,542	27	–23
Surveillance/target acquisition and tracking radar	17	5,370	6,123	27	–33
EOD/UDT	19	4,629	4,480	27	–25
Musicians, general	342	5,105	3,583	26	–17
Intercept operators (code and non-code)	33	5,454	1,213	26	–20
Not occupationally qualified, general	653	5,209	3,073	25	–17
Automatic data processing computers, general	438	5,497	4,781	25	–25
Image interpretation	261	6,847	3,819	25	–17
Special forces	174	8,034	11,878	25	–24
Dental care, general	196	7,782	6,476	24	–24
Surveying	186	6,136	5,322	23	–26
Auxiliaries	111	7,529	6,296	23	–24
Signal intelligence/electronic warfare, general	16	7,200	4,728	23	–25
Medical administration	308	6,729	5,805	23	–22
Air crew, general	252	6,490	7,430	23	–22
Surgery	130	7,897	7,169	22	–24

Table A.7 (continued)

Three-Digit Occupation	Number of Observations	Earnings Difference		Any Loss, %	
		Gross	Net	Gross	Net
Aircraft structures	315	5,579	5,847	22	–30
Radio/radar, general	379	7,278	5,883	22	–26
Missile guidance and control	160	6,884	5,131	22	–19
Artillery repair	33	12,009	12,488	22	–28
Fabric, leather, and rubber, general	155	5,310	5,502	21	–34
Veterinary medicine	60	9,758	7,966	21	–18
Boatswains	277	9,472	9,039	21	–28
Sonar, general	38	8,291	7,081	21	–27
Main propulsion	62	7,330	6,433	21	–27
Aircraft engines	697	6,341	7,216	21	–35
Electricians	445	9,206	8,773	21	–29
Nuclear weapons equipment repair, general	69	10,618	9,646	21	–26
Steelworking	115	8,581	8,552	21	–30
Construction, general	955	8,625	8,919	21	–33
Transportation	1,368	9,415	9,213	21	–30
Medical logistics	190	9,217	8,000	21	–25
B. Officers					
Physicians	573	5,029	565	38	–11
Other fixed-wing pilots	234	2,978	9,134	38	–16
Research and development coordinators	8	2,106	536	38	–18
Communications intelligence	11	4,617	2,645	38	–12
Fixed-wing fighter and bomber pilots	183	5,885	11,947	32	–23
Physical scientists	178	18,275	11,773	31	–21
Meteorologists	24	10,118	9,776	29	20
Executives, N.E.C.	77	17,616	33,642	26	–35
Administrators, general	18	18,654	20,842	25	–38
Nurses	882	11,229	12,754	25	–30
Dentists	170	14,546	16,526	23	–37
Supply	94	16,612	17,240	21	–29
Biomedical sciences and allied health officers	410	11,089	11,611	21	–33
Aviation maintenance and allied maintenance officers	130	8,758	8,027	21	–25

NOTES: Samples are restricted to reservists serving 0–30 active-duty days in 2004 and 31 or more active-duty days in 2005. All figures reported in $2004.

Exact Wording of 2004 and 2005 SOFS-R Earnings Questions

2004	2005
96. How much was your average monthly military compensation prior to your most recent activation, before taxes or other deductions? You can enter an amount here: Or, if you prefer, you can enter a range here. My average monthly military compensation was at least: but no more than:	96. How much was your average monthly military compensation (excluding reenlistment bonuses) in the 12 months prior to your most recent activation, before taxes and other deductions (i.e., gross pay)? You can enter an amount here: Or, if you prefer, you can enter a range here. My average monthly military compensation was at least: but no more than:
97. How much was your average monthly military compensation during your most recent activation, before taxes or other deductions? You can enter an amount here: Or, if you prefer, you can enter a range here. My average monthly military compensation was at least: but no more than:	97. How much was your average monthly military compensation (excluding reenlistment bonuses and imminent danger/hostile fire pay) during your most recent activation, before taxes and other deductions (i.e., gross pay)? You can enter an amount here: Or, if you prefer, you can enter a range here. My average monthly military compensation was at least: but no more than:
99. How much was your average monthly civilian income from all sources prior to your most recent activation, before taxes or other deductions? You can enter an amount here: Or, if you prefer, you can enter a range here. My average monthly civilian income was at least: but no more than:	102. How much were your average monthly civilian earnings from employment in the 12 months prior to your most recent activation, before taxes and other deductions (i.e., gross pay)? EXCLUDE income from bonuses, stocks and bonds, paid up life insurance, IRAs, savings, annuities, estate and trust payments, and rental income from property. You can enter an amount here: Or, if you prefer, you can enter a range here. My average monthly civilian earnings were at least: but no more than:
100. Did you have any civilian income during your most recent activation? Yes No	105. Did you have any civilian earnings from employment during your most recent activation? Yes No
101. How much was your average monthly civilian income from all sources during your most recent activation, before taxes or other deductions? You can enter an amount here: Or, if you prefer, you can enter a range here. My average monthly civilian income was at least: but no more than:	108. How much were your average monthly civilian earnings from employment during your most recent activation, before taxes and other deductions (i.e., gross pay)? EXCLUDE income from bonuses, stocks and bonds, paid up life insurance, IRAs, savings, annuities, estate and trust payments, and rental income from property. You can enter an amount here: Or, if you prefer, you can enter a range here. My average monthly civilian earnings were at least: but no more than:

APPENDIX C

Detailed Analysis of Differences in Military Earnings

The 2004 results in Table 3.3 strongly suggest that respondents systematically underreport their military earnings both before and during activation. The nature of this measurement error is explored in this appendix.

The Distribution of the Difference Between SOFS-R and Administrative Military Earnings Data

Tables C.1 and C.2 report summary statistics for individual difference between the SOFS-R and administrative estimates of military earnings. For pre-activation earnings, the median difference between the administrative data and survey data is smaller than the mean difference when using the month before or the 12 months before definitions. In contrast, the median is larger than the mean when months from previous activations are excluded from the calculations that use the administrative data.[1] The standard deviation of the estimated measurement error in military earnings is quite large. In fact, the null hypothesis that the average measurement error is zero cannot be rejected when months that were part of a prior activation are excluded from the calculation.[2] Nonetheless, underreporting of military earnings is widespread. Military earnings are higher in the administrative data than in the 2004 SOFS-R for between 62 and 78 percent of observations in the matched sample.[3]

Turning to the results for earnings during activation, the results in Table C.1 indicate that the average difference is larger than the median difference. This suggests that in a few observations, earnings in the administrative records are much higher than reported in the survey. In fact, monthly military earnings were at least $1,300 larger in the administrative data than in

[1] This pattern is consistent with prior activations contaminating the administrative data estimates that use the first two definitions for pre-activation earnings. This type of contamination would lead to very large administrative data estimates of pre-activation earnings for a small number of observations, which would drive up the mean discrepancy between the survey and administrative data but have a comparatively smaller effect on the median.

[2] Recall that months that were part of a prior activation are excluded to isolate pre-activation military earnings received while not activated. The standard error of the mean can be calculated by dividing the standard deviation by the square root of the sample size. For the estimate that excludes months from a prior activation from the calculations, the standard error is 106.

[3] With symmetric misreporting, earnings in the pay records should be greater than the survey response 50 percent of the time. This hypothesis is easily rejected by the data.

Table C.1

Distribution of Difference in Military Earnings, 2004

	Median	Mean	Std. Dev.	% (Administrative > Survey)	Mean Difference/ Administrative	Median Difference/ Administrative	Number
Pre-Activation							
Administrative Definition: Month Before Activation							
All	66.0	303.3	10,482.6	62	0.32	0.07	7,720
E1–E4	36.0	1.4	9,241.1	56	0.00	0.08	1,442
E5–E9	92.0	407.0	9,992.8	66	0.41	0.09	2,948
O1–O3	137.0	324.9	11,561.2	62	0.23	0.10	975
O4–O6	202.0	760.5	11,738.4	65	0.35	0.09	2,004
W1–W5	238.0	743.6	5,993.7	72	0.44	0.14	351
Administrative Definition: Average over 12 Months Before Activation							
All	187.5	274.2	9,327.9	78	0.30	0.21	7,720
E1–E4	108.3	37.5	8,767.8	71	0.08	0.22	1,442
E5–E9	211.6	344.9	8,631.8	80	0.38	0.23	2,948
O1–O3	341.9	274.1	10,246.6	78	0.20	0.25	975
O4–O6	455.9	731.1	10,541.6	83	0.34	0.21	2,004
W1–W5	371.9	455.1	4,748.9	85	0.32	0.26	351
Administrative Definition: Average over 12 Months Before Activation, Excluding Earlier Activation							
All	150.3	107.8	9,117.7	74	0.15	0.20	7,275
E1–E4	95.0	–53.5	8,603.1	68	–0.13	0.23	1,384
E5–E9	170.8	160.7	8,145.0	77	0.22	0.23	2,779
O1–O3	298.5	107.2	10,444.9	77	0.09	0.25	915
O4–O6	368.3	401.9	10,540.7	79	0.23	0.21	1,872
W1–W5	351.7	278.5	4,575.1	84	0.22	0.28	325
During Activation							
Unconditional							
All	334.7	530.7	10,697.9	65	0.15	0.09	7,720
E1–E4	241.9	346.3	12,031.2	63	0.14	0.10	1,442
E5–E9	363.7	589.4	10,597.2	64	0.16	0.10	2,948
O1–O3	727.7	528.5	11,426.5	72	0.10	0.14	975
O4–O6	868.6	899.7	10,005.9	74	0.12	0.12	2,004
W1–W5	400.5	398.7	5,545.7	65	0.08	0.08	351
Conditional on Administrative Data Reporting Military Earnings > 0							
All	462.7	788.2	9,920.8	72	0.20	0.12	7,258
E1–E4	326.4	538.6	11,592.8	69	0.21	0.12	1,345
E5–E9	474.7	818.9	9,659.0	72	0.21	0.12	2,767
O1–O3	996.8	1,136.1	10,788.3	83	0.20	0.17	906
O4–O6	1,052.3	1,330.2	8,921.1	80	0.17	0.13	1,941
W1–W5	641.5	899.9	4,130.9	72	0.15	0.11	299

Table C.2

Distribution of Difference in Military Earnings, 2005

	Median	Mean	Std. Dev.	% (Administrative > Survey)	Mean Difference/ Administrative	Median Difference/ Administrative	Number
				Pre-Activation			
				Administrative Definition: Month Before Activation			
All	–25.0	–3,031.1	173,154.5	45	–3.12	–0.03	33,296
E1–E4	–37.0	–2,016.3	112,301.7	43	–3.73	–0.07	5,826
E5–E9	–22.0	–3,769.2	225,399.8	46	–4.05	–0.02	17,860
O1–O3	–34.0	–2,241.8	44,987.3	47	–1.38	–0.02	3,212
O4–O6	49.0	–2,618.3	41,725.6	53	–1.05	0.02	5,503
W1–W5	11.0	–4,062.5	37,375.4	51	–2.36	0.01	895
				Administrative Definition: Average over 12 Months Before Activation			
All	66.6	–3,137.3	173,118.7	56	–3.62	0.08	33,296
E1–E4	21.3	–2,098.9	112,236.2	51	–4.59	0.05	5,826
E5–E9	80.3	–3,852.5	225,378.6	57	–4.54	0.09	17,860
O1–O3	146.9	–2,444.0	44,858.1	58	–1.72	0.10	3,212
O4–O6	310.6	–2,895.8	41,453.1	65	–1.30	0.14	5,503
W1–W5	241.0	–4,291.0	37,500.2	64	–2.87	0.16	895
				Administrative Definition: Average over 12 Months Before Activation, Excluding Earlier Activation			
All	49.3	–3,283.8	176,852.8	54	–4.49	0.07	31,853
E1–E4	13.8	–2,149.2	113,264.8	50	–5.11	0.03	5,719
E5–E9	58.5	–4,034.3	230,649.4	55	–5.77	0.08	17,028
O1–O3	114.2	–2,626.0	45,629.4	57	–2.08	0.09	3,085
O4–O6	247.6	–3,276.3	42,324.2	63	–1.75	0.13	5,152
W1–W5	229.5	–4,456.9	37,854.6	63	–3.30	0.17	869
				During Activation			
				Unconditional			
All	227.5	–2,407.9	158,148.7	58	–0.67	0.06	33,296
E1–E4	83.1	–2,364.2	251,213.4	52	–1.02	0.04	5,826
E5–E9	270.6	–2,393.0	158,245.5	60	–0.67	0.08	17,860
O1–O3	507.9	–2,384.9	47,768.7	63	–0.49	0.10	3,212
O4–O6	855.5	–2,436.1	40,133.7	69	–0.32	0.11	5,503
W1–W5	383.6	–4,142.9	43,871.4	62	–0.76	0.07	895
				Conditional on Administrative Data Reporting Military Earnings > 0			
All	350.2	–1,967.3	129,484.1	64	–0.51	0.09	32,197
E1–E4	197.8	–1,423.5	116,837.7	59	–0.55	0.08	5,501
E5–E9	349.8	–2,247.2	159,972.8	64	–0.60	0.09	17,468
O1–O3	803.0	–1,713.2	49,404.5	76	–0.30	0.14	2,957
O4–O6	1,024.9	–2,137.6	40,301.9	73	–0.27	0.13	5,425
W1–W5	539.0	–3,594.3	44,280.0	67	–0.61	0.09	846

the survey for 25 percent of reservists in the matched sample. Although the standard deviation of the difference in observed military earnings is large, the hypothesis that the average discrepancy is zero can be rejected at the 1 percent level. As with pre-activation military earnings, underreporting appears to be widespread; between 65 and 72 percent of survey respondents report less military earnings than indicated by the administrative data.

Table C.1 also shows the mean and median difference between estimates of military earnings in the administrative and survey data by pay grade. These results indicate that the absolute size of underreporting increases with pay grade. However, this may simply reflect the fact that military pay increases substantially with seniority. Columns 5 and 6 normalize the median and mean measurement error in military earnings by the estimate of military earnings derived from the administrative data. For the pre-activation period, mean measurement error relative to average military earnings is larger for more senior reservists (for both officers and enlisted reservists). However, this is not the case for median measurement error. Furthermore, mean and median measurement error of earnings during activation normalized by average earnings does not appear to be systematically related to pay grade. Overall, these findings suggest that misreporting of military earnings as a percentage of "true" earnings is not strongly related to pay grade.

Are Respondents Reporting Only Basic Pay?

In the literature examining the quality of income data from surveys, a consistent finding is that respondents report lower amounts when asked a single question about income rather than a battery of questions about income from various sources (Bound, Brown, and Mathiowetz, 2001; Nelson et al., 1998; Roemer, 2000). It is conjectured that this occurs because respondents do not remember all of their sources of income unless specifically prompted about each one. In particular, respondents tend to underreport income from sources that contribute a smaller share to their total income or from sources where payments are made irregularly.

This phenomenon might explain why military income is underreported in the SOFS-R. Military compensation comes not only in the form of basic pay but also in bonuses, special pays, and in a variety of allowances. Thus, the underreporting we observe may be due to respondents reporting only their primary component of military income, basic pay.

For most respondents, this does not appear to be the case. As seen in Table C.3, the average and median difference between basic pay and the survey response for the 2004 survey is negative for both during- and pre-activation earnings.[4] This pattern also holds across all pay grades. However, it should be noted that a nontrivial fraction of respondents, about 13 percent, report earnings during activation that are within 10 percent of average monthly basic pay (according to the administrative data) received during activation.[5] Thus, it is possible that some respondents are underreporting in this fashion, although there is no way to definitively prove that conjecture.

---

[4] This pattern also holds in the 2005 survey, but as discussed above, the 2005 earnings data are subject to severe overreporting among some respondents.

[5] In comparison, about 30 percent report earnings that are within 10 percent of what the administrative data indicate are total military earnings.

Table C.3
Distribution of Difference in Military Earnings: Basic Pay, 2004

	Median	Mean	Std. Dev.	% (Administrative > Survey)	Number
Pre-Activation					
Administrative Definition: Month Before Activation					
All	–216.0	–161.9	9,668.1	24	7,720
E1–E4	–180.0	–241.7	8,769.5	19	1,442
E5–E9	–250.0	–70.5	8,894.4	26	2,948
O1–O3	–400.0	–348.1	10,968.5	25	975
O4–O6	–650.0	–324.5	11,116.4	25	2,004
W1–W5	–402.5	–112.7	5,487.4	27	351
Administrative Definition: Average over 12 Months Before Activation					
All	–123.3	–159.4	8,832.7	29	7,720
E1–E4	–95.9	–191.9	8,505.4	27	1,442
E5–E9	–128.8	–103.8	7,854.2	30	2,948
O1–O3	–212.5	–344.9	10,078.3	31	975
O4–O6	–338.5	–260.9	10,214.9	28	2,004
W1–W5	–215.6	–262.4	4,534.0	29	351
Administrative Definition: Average over 12 Months Before Activation, Excluding Earlier Activation					
All	–145.2	–281.2	8,842.7	22	7,275
E1–E4	–112.3	–253.2	8,518.1	21	1,384
E5–E9	–155.8	–239.8	7,675.3	22	2,779
O1–O3	–237.8	–474.1	10,330.6	27	915
O4–O6	–416.1	–530.4	10,314.5	20	1,872
W1–W5	–219.3	–416.2	4,458.6	24	325
During Activation					
All	–708.3	–640.6	9,430.2	26	7,720
E1–E4	–576.0	–532.2	11,527.0	26	1,442
E5–E9	–729.5	–594.5	9,558.0	27	2,948
O1–O3	–787.0	–923.9	10,175.0	24	975
O4–O6	–995.0	–1,125.7	7,407.2	25	2,004
W1–W5	–1,185.4	–1,148.3	4,813.4	23	351
Conditional on Administrative Data Reporting Military Earnings > 0					
All	–588.0	–443.4	8,886.6	29	7,258
E1–E4	–472.0	–391.8	11,460.6	28	1,345
E5–E9	–620.9	–415.4	8,998.5	30	2,767
O1–O3	–541.2	–477.0	9,869.3	30	906
O4–O6	–798.0	–758.3	6,313.9	28	1,941
W1–W5	–988.6	–775.1	4,031.5	25	299

Bibliography

Baum-Snow, Nathaniel, and Derek Neal, "Mismeasurement of Usual Hours Worked and Its Impact on Measured Wages and Wage Gaps Over Time," *Review of Economics and Statistics* (forthcoming, 2008).

Baj, John, Charles Trott, and David Stevens, "A Feasibility Study of the Use of Unemployment Insurance Wage-Record Data as an Evaluation Tool for JTPA: Report on Project Phase I Activities," National Commission on Employment Policy, Research Report 90-02, 1991.

Bethlehem, Jelke, "Weighting Nonresponse Adjustments Based on Auxiliary Information," in Robert M. Groves, Don A. Dillman, John L. Eltinge, and Roderick J.A. Little, eds., *Survey Nonresponse*, New York: Wiley, 2002, pp. 275–288.

Blakemore, Arthur, Paul Burgess, Stuart Low, and Robert St. Louis, "Employer Tax Evasion in the Unemployment Insurance Program," *Journal of Labor Economics*, Vol. 14, 1996, pp. 210–230.

Bollinger, Christopher, "Measurement Error in the Current Population Survey: A Nonparametric Look," *Journal of Labor Economics*, Vol. 16, No. 3, 1998, pp. 576–594.

Bolstein, Richard, "Comparison of the Likelihood to Vote Among Preelection Poll Respondents and Nonrespondents," *Public Opinion Quarterly*, Vol. 55, 1991, pp. 648–650.

Bound, John, and Alan Krueger, "The Extent of Measurement Error in Longitudinal Earnings Data: Do Two Wrongs Make a Right?" *Journal of Labor Economics*, Vol. 9, 1991, pp. 1–24.

Bound, John, Charles Brown, and Nancy Mathiowetz, "Measurement Error in Survey Data," in James Heckman and Edward Leamer, eds., *Handbook of Econometrics*, Vol. 5, Elsevier Science, 2001, pp. 3705–3842.

Conover, William Jay, *Practical Nonparametric Statistics*, 3rd Edition, New York: Wiley, 1999.

Curtin, Richard, Stanley Presser, and Elanor Singer, "The Effects of Response Rate Changes on the Index of Consumer Sentiment," *Public Opinion Quarterly*, Vol. 64, 2000, pp. 413–428.

Defense Manpower Data Center, "May 2004 Status of Forces Survey of Reserve Component Members: Administration, Datasets, and Codebook," DMDC Report No. 2004-013, Washington, D.C., 2004.

———, "June 2005 Status of Forces Survey of Reserve Component Members: Administration, Datasets, and Codebook," DMDC Report No. 2005-023, Washington, D.C., 2005.

Denmead, Gabrielle, and Joan Turek, "Comparisons of Health Indicators by Income in Three Major Surveys," Papers and Proceedings of the American Statistical Association, Minneapolis, Minn., 2005.

Goldman, Dana, and James Smith, "Commentary: Methodological Biases in Estimating the Burden of Out-of-Pocket Expenses," *Health Services Research*, Vol. 35, No. 6, 2001, pp. 1357–1370.

Groves, Robert, "Nonresponse Rates and Nonresponse Bias in Household Surveys," *Public Opinion Quarterly*, Vol. 70, No. 5, 2006, pp. 646–675.

Haider, Steven J., and David S. Loughran, "The Effect of the Social Security Earnings Test on Male Labor Supply: New Evidence from Survey and Administrative Data," *Journal of Human Resources*, Vol. 43, No. 1, 2008, pp. 57–87.

Hill, Carolyn J., V. Joseph Hotz, Charles H. Mullin, and John Karl Scholz, "EITC Eligibility, Participation and Compliance Rates for AFDC Households: Evidence from the California Caseload," Chicago, Ill.: Joint Center for Poverty Research, working paper 102, July 1, 1999.

Hotz, Joseph, and John Karl Scholz, "Measuring Employment and Income Outcomes for Low-Income Populations with Administrative and Survey Data," in *Studies of Welfare Populations: Data Collection and Research Issues,* National Research Council, Washington, D.C.: National Academy Press, 2002, pp. 275–315.

Hurd, Michael D., and Susann Rohwedder, "Economic Well-Being at Older Ages: Income- and Consumption-Based Measures in the HRS," Santa Monica, Calif.: RAND Corporation, WR-410, 2006. As of March 24, 2008:
http://www.rand.org/pubs/working_papers/WR410/

Jacobson, Louis, Robert LaLonde, and Daniel Sullivan, "Earnings Losses of Displaced Workers," *American Economic Review*, Vol. 83, No. 4, 1993, pp. 685–709.

Kapteyn, Arie, and Jelmer Y. Ypma, "Measurement Error and Misclassification: A Comparison of Survey and Administrative Data," *Journal of Labor Economics,* Vol. 25, No. 3, 2007, pp. 513–551.

Keeter, Scott, Carolyn Miller, Andrew Kohut, Robert Groves, and Stanley Presser, "Consequences of Reducing Nonresponse in a National Telephone Survey," *Public Opinion Quarterly*, Vol. 64, 2000, pp. 125–148.

Loughran, David S., Jacob Alex Klerman, and Craig W. Martin, *Activation and the Earnings of Reservists,* Santa Monica, Calif.: RAND Corporation, MG-474-OSD, 2006. As of March 24, 2008:
http://www.rand.org/pubs/monographs/MG474/

Loughran, David S., Jacob Alex Klerman, and Bogdan Savych, *The Effect of Reserve Activations and Active Duty Deployments on Local Employment During the Global War on Terrorism,* Santa Monica, Calif.: RAND Corporation, TR-321-RC, 2006. As of March 24, 2008:
http://www.rand.org/pubs/technical_reports/TR321/

Merkle, Daniel, and Murray Edelman, "Nonresponse in Exit Polls. A Comprehensive Analysis," in R. Groves, D. Dillman, J. Eltinge, and R. J. Little, eds., *Survey Nonresponse,* New York: Wiley, 2002, pp. 243–357.

Merkle, Daniel, Murray Edelman, Kathy Dykeman, and Chris Brogan, "An Experimental Study of Ways to Increase Exit Poll Response Rates and Reduce Survey Error," paper presented at the annual meeting of the American Association for Public Opinion Research, St. Louis, Mo., 1998.

Nelson, Charles, Marc Roemer, Daniel Weinberg, and Edward Welniak, "Fifty Years of United States Income Data from the Current Population Survey, Housing and Household Economic Statistics Division," U.S. Bureau of the Census, paper prepared for the Federal Committee on Statistical Methodology meeting, Washington, D.C., November 1998.

Office of the Under Secretary of Defense (Comptroller), National Defense Budget Estimates for FY 2006 [the Green Book], Washington, D.C., April 2005. As of March 24, 2008:
http://www.defenselink.mil/comptroller/defbudget/fy2006/fy2006_greenbook.pdf

Olson, Kristen, "Survey Participation, Nonresponse Bias, Measurement Error Bias, and Total Bias," *Public Opinion Quarterly,* Vol. 70, No. 5, 2006, pp. 737–758.

Rodgers, Willard L., Charles Brown, and Gregory Duncan, "Errors in Survey Reports of Earnings, Hours Worked, and Hourly Wages," *Journal of the American Statistical Association*, Vol. 88, 1993, pp. 1208–1218.

Roemer, Marc, "Assessing the Quality of the March Current Population Survey and the Survey of Income and Program Participation Income Estimates, 1990–1996," Mimeo, Income Statistics Branch, U.S. Bureau of the Census, 2000.

Schochet, Peter, Sheena McConnell, and John Burghardt, "National Job Corps Study: Findings Using Administrative Earnings Records Data," Princeton, N.J.: Mathematica Reference Number 8140-840, 2003.